Introduction

Welcome to AS Guru™ Biology!

Studying AS Biology will build on your experience and understanding of biology that you have studied up to GCSE. In order to get the most out of this course you will need to keep asking yourself what you already know and, more importantly, work out what you don't know. Good note-taking skills are essential and you need to be organised and take responsibility for your own learning. Keep questioning yourself and others about the material.

This book covers the main areas of your course and the material will help you to pass examinations in the subject at C grade. If you think that you are capable of higher grades, use this book as a basis for understanding the main points and then read round the subject matter in more detail.

There is no need to read the book from cover to cover. Simply dip into the parts that are relevant to you at any time. If you need to know something in order to understand a particular section you will be told about it in the text.

Each section starts with an introduction that outlines the main areas to be covered. You will be reminded of the important concepts and details from your KS4 work, which you will need to remember in order to make the most of the material. You will find that each section develops logically, from simple to more complex ideas. There is also a summary that covers the key concepts in the section, sample questions and suggested demonstrations of your key skills (see page 6).

It is important at this level of study that you are able to understand and use the appropriate level of terminology, correctly. To help you, important words and phrases are highlighted in bold throughout the text, and you can find brief definitions of them in the glossary at the back of the book.

Need a little more help?

This book is part of a multimedia educational service offering support and is designed to complement two other major resources, which will help you with your biology studies – the AS Guru™ website and AS Guru™ TV.

Visit the AS Guru™ website at www.bbc.co.uk/education/asguru/biology for up-to-date information, interactive tutorials and tests, and details of AS Guru™ TV. Check your listings magazine for Videoplus codes so that you don't miss the programmes.

Course structure

AS Biology consists of three modules, which are assessed by examination. You will also have to complete a piece of coursework or sit an externally assessed practical examination, depending on your examination board and study centre. In both cases, you will need to demonstrate your practical and investigative skills.

If you go on to study A2 Biology, you will carry out work in three further modules. At the end of the A2 year, there is also a synoptic examination paper where you will need to bring together all the work you have covered over the two years.

The content for each examination board is summarised on the opposite page. If there is a particular part of the book that is not required by all the boards, you will see the logo of the boards that do require it, in the margin.

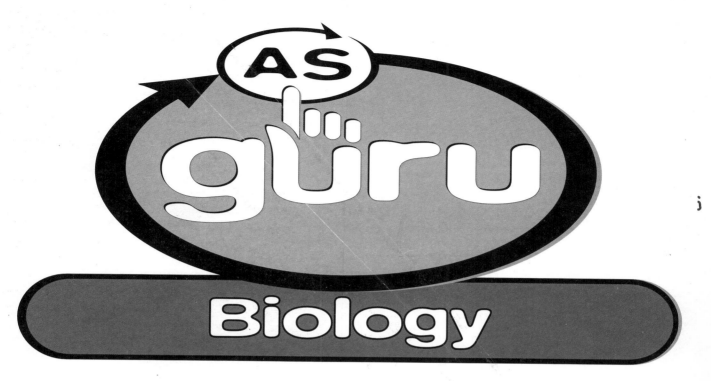

John Graham and Anthony Lewis

Key skills
Key skills information p6,
taken from the QCA website (www.qca.org.uk).

Picture credits
Biophoto Associates: p8 (lower margin); p94.

Oxford Scientific Films: / Science Picture p8 (upper margin);
/ London Scientific Films p8 (top); / Animals Animals Catalogue/David
Dennis p38; / Waina Cheng p135 (top);
/ Breck P Kent p135 (bottom); / George Bernard p143.

Science Photo Library: / Professors P Motta & T. Naguro p15 (all);
/ D. Phillips p16 (top); / CNRI p16 (middle); / Alfred Pasieka p16
(bottom); / Pr S Cinti/Universite D'Ancone / CNRI p17 (top); / Biology
media p17 (middle); / Bill Loncore p17 (bottom); / National Institute of
Health p26 (top); / Ken Edward p26 (bottom); / Manfred Kage p31; /
Hermann Eisenbeiss p38 (bottom left); / Sinclair Stammers p38 (right);
/ Biology Media p54 (left); / Eye of Science p54 (right); / Sinclair
Stammers p80; / Phillipe Plailly p81; / CNRI p97; / Martin Dohrn/Royal
College of Surgeons p104; / Juergen Berger, Max-Planck Institute
p108; / Dr Jeremy Burgess p118; / Dr Morely Read p126; / Andrew
Syred p128; / Rod Planck p129; / Alfred Pasieka p135.

Still Pictures: / Klaus Andrew p143 (bottom).

Published by BBC Educational Publishing, BBC White City,
201 Wood Lane, London W12 7TS.

First published 2001

Colour reproduction, printed and bound by sterling press,
Northamptonshire.

To place an order, please telephone 87 541001 (Monday – Friday, 0800 – 1800)
or write to BBC Education, PO Box 234, Wetherby, West Yorkshire, LS23 7EU.

Visit the BBC Education website at: www.bbc.co.uk/education

Contents

Topic	EDEXCEL	AQA A	AQA B	OCR
Prokaryotic cells	✔	✔	✔	✔
Eukaryotic cell	✔	✔	✔	✔
Plasma membranes and cell transport	✔	✔	✔	✔
Microscopes	✔	✔	✔	✔
Cell fractionation		✔	✔	
Carbohydrates	✔	✔	✔	✔
Proteins	✔	✔	✔	✔
Lipids	✔	✔	✔	✔
Structure of nucleic acids	✔	✔	✔	✔
Water	✔	✔	✔	✔
Enzymes	✔	✔	✔	✔
Tissues	✔	✔		✔
Organ histology	✔	✔		✔
Tissue fluid	✔		✔	
Heart and circulation	✔	✔	✔	✔
Transport in plants	✔		✔	✔
Xerophytes	✔		✔	✔
Human respiratory system	✔		✔	✔
Gas exchange in flowering plants	✔		✔	✔
Cell cycle, mitosis, roles of meiosis	✔	✔	✔	✔
DNA, RNA, replication and protein synthesis	✔	✔	✔	✔
Recombinant DNA techniques		✔	✔	✔
Genetic fingerprinting		✔	✔	
Polymerase chain reaction		✔	✔	
Enzyme biotechnology	✔	✔		
Human reproduction and birth	✔	✔		
Human digestive system	✔		✔	
Human diet	✔			✔
Malnutrition				✔
Mineral nutrition in animals and plants	✔			✔
Hydrophytes	✔			✔
Gas transport in blood	✔			✔
Ecosystems, energy flow and recycling of elements	✔			✔
Immunity – B-cells		✔		✔
Immunity – T-cells and allergies				✔
Exercise effects on the body		✔		✔
Gas exchange surfaces – protozoa, gills in fish	✔		✔	
Reproduction in flowering plants	✔			
Special modes of nutrition	✔		✔	
Management of energy resources	✔			
Human influences on the environment	✔			
Cultivated plants, cereals		✔		
Herbicides and pesticide		✔		
Gene therapy			✔	
Chromatography		✔	✔	
Mutation			✔	
Smoking-related diseases				✔
Infectious diseases				✔
Human genome project	✔			✔
Structure of viruses			✔	
Food chains and succession			✔	
Populations			✔	

Your study centre may suggest that you sit your first module assessment in January, in order to give you experience and to get one mark 'under your belt'. If you are disappointed with the result and think you could do better, don't panic! You can re-sit the exam once and the highest mark is the one that counts.

Key skills

You'll find ways to demonstrate your key skills throughout the book, and there are ideas and examples in the section summaries.

Your work at AS or A2 level can help you gain a new, key skills qualification. This award demonstrates that you have developed important life skills – they are in demand by employers and Further Education establishments. To achieve a key skills qualification, you need to collect a portfolio demonstrating your level of competence in six areas. There are three levels of award and you should aim to meet the criteria laid down for level 3 (see below). Anytime you are involved in a class discussion, writing an essay, designing and carrying out an experiment, making a presentation, researching or carrying out a project, you have the opportunity to add a copy to your portfolio. The table below summarises the criteria for each key skill.

C3	**Communication level 3**
C3.1a	Contribute to a group discussion about a complex subject.
C3.1b	Make a presentation about a complex subject, using at least one image to illustrate complex points.
C3.2	Read and synthesise information from two extended documents that deal with a complex subject. One of these documents should include at least one image.
C3.3	Write two different types of documents about complex subjects. One piece of writing should be an extended document and include at least one image.
N3	**Application of number level 3**
N3.1	Plan, and interpret information from two types of sources including a large data set.
N3.2	Carry out multi-stage calculations to do with: amounts and sizes; scales and proportion; handling statistics and rearranging and using formulae. Work with a large data set on at least one occasion.
N3.3	Interpret results of your calculations, present your findings and justify your methods. You must use at least one graph, one chart and one diagram.
IT3	**IT level 3**
IT3.1	Plan and use different sources to search for, and select, information required for two different purposes.
IT3.2	Explore, develop and exchange information and derive new information to meet two different purposes.
IT3.3	Present information from different sources for two different purposes and audiences. Your work must include at least one example of text, one example of an image and one example of numbers.
WO3	**Working with others level 3**
WO3.1	Plan complex work with others, agreeing objectives, responsibilities and working arrangements.
WO3.2	Seek to establish and maintain co-operative working relationships over an extended period of time, agreeing changes to achieve agreed objectives.
WO3.3	Review work with others and agree ways of improving collaborative work in the future.
LP3	**Learning performance level 3**
LP3.1	Agree targets and plan how these will be met over an extended period of time, using support from appropriate people.
LP3.2	Take responsibility for your learning by using your plan and seeking feedback and support from relevant sources to help meet targets.
LP3.3	Review progress on two occasions and establish evidence of achievements, including how you have used learning from other tasks to meet new demands.
PS3	**Problem solving level 3**
PS3.1	Explore a complex problem, coming up with three options for solving it and justify the option selected for taking forward.
PS3.2	Plan and implement at least one option for solving the problem, review progress and revise your approach as necessary.
PS3.3	Apply agreed methods to check if the problem has been solved, describe the results and review your approach to problem solving.

Cell structure, function and organisation

In this section you will be learning about:

- ☞ light and electron microscopes and the difference between magnification and resolution
- ☞ interpreting drawings and photographs of cells organelles
- ☞ how to compare and contrast plant and animal cells
- ☞ prokaryotic and eukaryotic cells.

This section looks at the structure and function of organelles found in plant and animal cells. It is important that you understand how individual cells relate to the whole organism. While there are many **unicellular** (single-celled) organisms, such as bacteria and amoeba, most organisms are **multicellular**.

Multicellular organisms have a distinct advantage by being insulated against their environment. Although they have the same basic cellular structure, cells in a multicellular organism are **differentiated** – a process that creates specific cellular features to play a particular role within the organism. In the reproduction and exchange and transport sections, you will come across many types of differentiated cells and their role in the life of their organism.

In KS4 you learnt that **cells** are organised into **tissues**, tissues into **organs**, organs into **organ systems** and organ systems into **organisms**. At this highest level of organisation, there is a division of labour between cells so that they perform quite specific tasks at different levels of organisation within the organism.

Remember:

- a tissue is a collection of differentiated cells (and any products produced by them), which are specialised for a particular function within the organism

- tissue may be formed by thesame or mixed types of cell and the study of tissues is called histology

- organ systems consist of two or more organs working together to provide a major function or process within the organism.

Light microscopy

You will probably have used a light microscope at KS4. The cells you looked at may have been from the lining of your cheek, or the **epidermis** of an onion. You will have looked at the difference between animal and plant cells and drawn them.

KEY SKILLS C3.1b

At AS level, you will be asked to make detailed drawings of specimens seen under the microscope. Avoid shading and use a pencil, rather than colour, in your diagrams. Always use clear, unbroken lines and show as much detail as you can. Label the details clearly so that everyone will know exactly which part you mean. You should also write down the **magnification** of your view and any measurements you have taken.

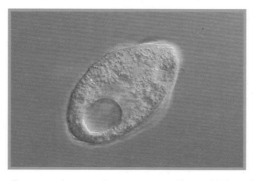
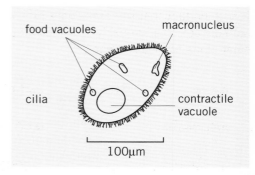

food vacuoles — macronucleus

cilia

contractile vacuole

100µm

Paramecium cell as seen under a light microscope and corresponding diagram

The light microscope

Light is passed through a very thin specimen placed on a slide and positioned on the microscope stage. The image is magnified using **eyepiece lenses** and an **objective**.

Tomato cell as seen under high power of a light microscope

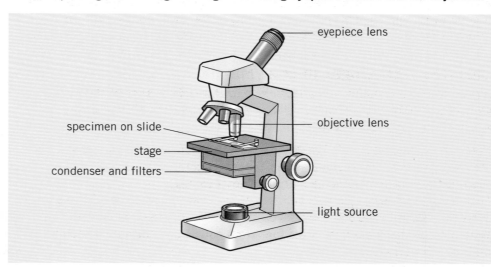

eyepiece lens

specimen on slide

objective lens

stage

condenser and filters

light source

Light microscope

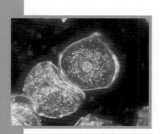

Tomato cell as seen under high power of a light microscope

Things to remember when using a light microscope

- Cells are generally colourless and may need to be stained in order to see their structure clearly. A typical stain is **methylene blue,** which colours nuclei blue.
- Specimens can dry up, so you may need to add a drop of water and use a cover-slip. Take care to avoid trapping air bubbles as these can be confused with cells.

- Specimens need to be one cell thick so that enough light can pass through, into the objective lens.
- Cells are three-dimensional objects and a microscope can only focus on a flat plane. When working under high power (highest magnification), you should use the fine focusing control to look at different levels through the specimen so that you can build up a three-dimensional impression.
- Always start by focusing your specimen under low power (usually the shortest objective lens with the smallest number), using the coarse focusing control. Then move the medium-powered lens into position and focus with the fine control. Repeat this procedure for the high-power objective lens.
- Always place the low-powered lens into position before removing a microscope slide. This will prevent the more high-powered lenses from getting scratched.

It is important that you appreciate just how small cells (and their structures) actually are. You are likely to come across questions where you will have to work in small units of measurement.

Name	Symbol	Part of a metre
millimetre	mm	$0.001 = 10^{-3}$ = a thousandth
micrometre	μm	$0.000001 = 10^{-6}$ = a millionth
nanometre	nm	$0.000000001 = 10^{-9}$ = a thousand millionth

Comparison of measurements used in microscopy

The human eye can just see objects sized around 100μm. This corresponds to the size of an **ovum** (human egg cell), which is unusually large. Typical cells are 5–20μm in diameter and structures within them vary from 0.5–1μm.

> **GURU TIP**
> To calculate the magnification, multiply the number etched on the side of the objective lens by the number etched onto the eyepiece lens.

Measuring with a light microscope

Using an **eyepiece graticule** and a **stage micrometer slide** (a special slide marked every 0.01mm), you will be able to measure the size of structures seen under the light microscope.

- Place an eyepiece graticule into the eyepiece lens (as you look through the eyepiece lens, you should see the graticule scale).
- Place a stage micrometer slide on the microscope stage.
- Look through the eyepiece lens and rotate it so that the eyepiece graticule scale lines up with the stage micrometer scale (you will need to focus).
- Work out how many eyepiece scale units (eu) equals one stage micrometer unit (su). For example:
  if 4eu = 1su, then 1eu = 0.01/4 = 0.0025mm = 2500μm.
- Replace the stage micrometer with the specimen slide (don't change either the eyepiece or objective lenses) and measure the required parts using eyepiece units.

> **Remember:** recalibrate your microscope every time you change the objective (or eyepiece) lens.

KEY SKILLS N3

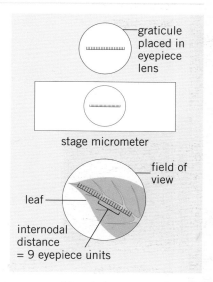

Measuring a specimen, using a graticule and stage micrometer slide

Microscopy: comparing cells

In KS4, you came across various structures found in cells, called **organelles**. These included the **nucleus**, **cytoplasm**, **cell membrane** and **mitochondria**, which are all found in plant and animal cells.

> **Remember: chloroplasts, cell walls** and large, permanent **vacuoles** are only found in plant cells.

KEY SKILLS
PS3, LP3

When you view specimens of plant and animal cells under a very high quality light microscope, you can see many other organelles that you will not have come across before. The diagrams show the various organelles that can be seen. You will look at these in more detail, later on.

GURU TIP
Don't panic!
Start by finding the familiar nucleus, cytoplasm, cell membrane and mitochondria.
Then look at each of the extra organelles in turn.

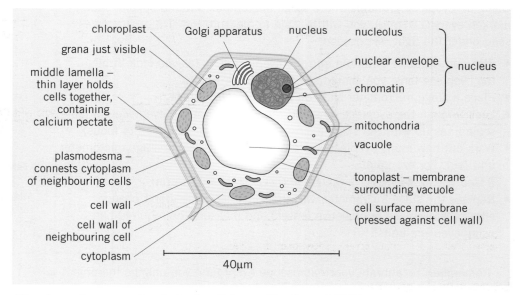

Structure of a generalised animal cell (magnification x1500)

Compare and contrast
Try and relate this drawaing to the generalised animal cell, above.

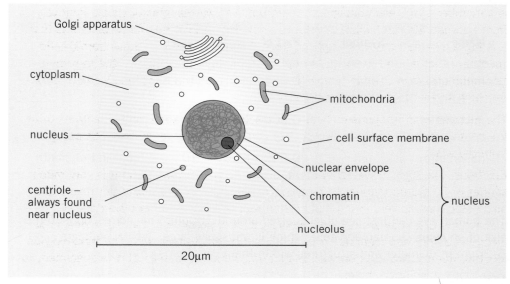

Structure of a generalised plant cell (magnification x1500)

Electron microscopy

Light microscopes have their limits and are unable to distinguish between small cellular structures. To help you to understand this, it's important to make the distinction between **magnification** and **resolution**. Inkjet printers print images using a pattern of dots on paper at a resolution of between 300 and 1200 dots per inch (dpi). If you took an image at 300dpi and magnified it 100 times, you would not be able to see any more detail than in the original because of the resolution.
If you compared this with a 1200dpi resolution and magnified it by the same amount, you would see more detail.

The principle of resolution is important to microscopy. The maximum, useful magnification of a light microscope is about 1500 times. Beyond this magnification, details cannot be seen and smaller objects cannot be resolved. This is to do with the nature of light.

In KS4 Physics, you learned that light is a form of energy and travels in waves. Light is a member of the electromagnetic spectrum. The wavelengths of visible light vary from between 400nm (violet) to 700nm (red). The general rule for resolution is that it is limited by half the wavelength of the radiation (as part of the electromagnetic spectrum) that is used to see it. For light and the light microscope, the maximum resolution occurs at 200nm (equivalent to approximately a magnification of 1500). Anything smaller than this can't be resolved.

The electron microscope uses beams of electrons rather than light. Electrons orbit the central nucleus of atoms and if they gain sufficient energy, they can escape from the atom.

Electrons, like light, travel in the form of waves, but the wavelength is very much shorter at 0.005nm (similar to X-rays). This means that the maximum resolution, using electrons, is very much higher than that of light. Another advantage of electrons is that they are negatively charged. It is possible to focus the beams of electrons using the magnetic fields produced by electromagnets. The **transmission electron microscope** uses beams of electrons, which pass through the specimen, to view a high resolution image.

> **GURU TIP**
> **Magnification and resolution**
> These are easy to get mixed up, so try and learn their definitions:
>
> **Magnification** is an expression of how much an object has been enlarged or diminished.
>
> **Resolution** is the ability to distinguish between two points. If two points lie close together and can't be seen, then they can't be resolved.

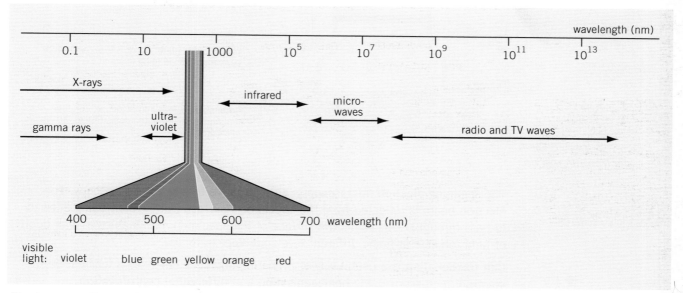

Electromagnetic spectrum

Electron microscopy and cell fractionation

GURU TV
Electron microscopy can tell you a lot about the structure of cell organelles. Cell specialist, Paul Nurse, looks into the options available for examining their structures.

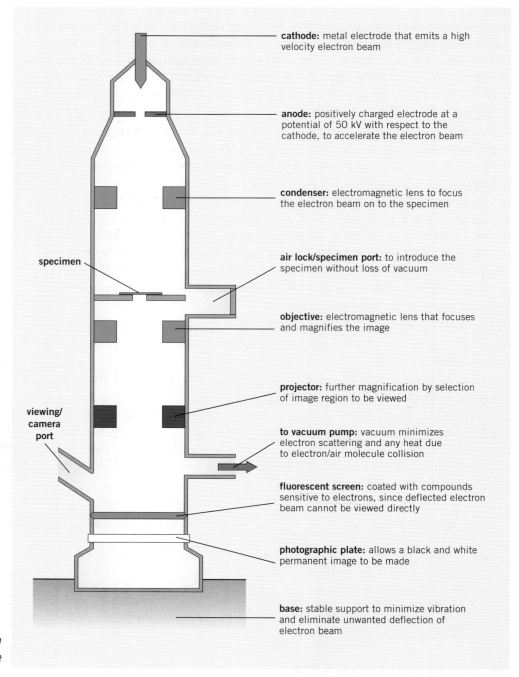

cathode: metal electrode that emits a high velocity electron beam

anode: positively charged electrode at a potential of 50 kV with respect to the cathode, to accelerate the electron beam

condenser: electromagnetic lens to focus the electron beam on to the specimen

specimen

air lock/specimen port: to introduce the specimen without loss of vacuum

objective: electromagnetic lens that focuses and magnifies the image

projector: further magnification by selection of image region to be viewed

viewing/
camera
port

to vacuum pump: vacuum minimizes electron scattering and any heat due to electron/air molecule collision

fluorescent screen: coated with compounds sensitive to electrons, since deflected electron beam cannot be viewed directly

photographic plate: allows a black and white permanent image to be made

base: stable support to minimize vibration and eliminate unwanted deflection of electron beam

Transmission electron microscope

The **scanning electron microscope** is an alternative to electron microscopy. The electron beam is reflected from the specimen's surface (rather than being allowed to pass through). This scans the surface of the specimen and enables a large depth of field (objects at many different levels are kept in focus).

Cell fractionation is a useful way of preparing organelle specimens for light or electron microscopy using **homogenisation** and **differential centrifugation**.

Homogenisation relies upon breaking up cells in a blender or pestle and mortar. This releases the organelles, many of which will not be damaged by the process.

···➔ AQA A
···➔ AQA B
···➔ WJEC

Differential centrifugation

Differential centrifugation uses the different densities of cell organelles, to separate them out for examination. If you shake up a sample of soil and water, the largest, most dense particles of soil will settle quickly to the bottom of the **homogenate**, forming a layer. Other particles of different size and density will settle on top of this, forming different layers. It is the force of gravity that pulls the particles down. This force can be increased, and the process speeded up, by placing samples in a centrifuge (basically a spinning top that turns thousands of times per minute).

1 A sample of the homogenate is spun at 700 times the force of gravity for 10 minutes inside centrifuge tubes (made of specially toughened glass to withstand the enormous stresses involved). The cell nuclei form a **sediment** on the bottom of the centrifuge tubes. Above the sediment are all the other organelles in solution. This is known as the **supernatant**.

2 The supernatant is removed and spun for a second time for 20 minutes at about 20 000 times the force of gravity. This forces mitochondria into the sediment.

3 The supernatant is removed again and spun for a third time for 60 minutes at about 105 000 times the force of gravity. Ribosomes, endoplasmic reticulum, lysosomes, centrioles and other organelles form the sediment. Proteins and the liquid making up the cytoplasm form the supernatant.

If plant tissues were being processed, chloroplasts would be found in the sediment at 700g

Cells

GURU TIP
Centrifugation is also a useful technique for separating out the four basic components of blood.

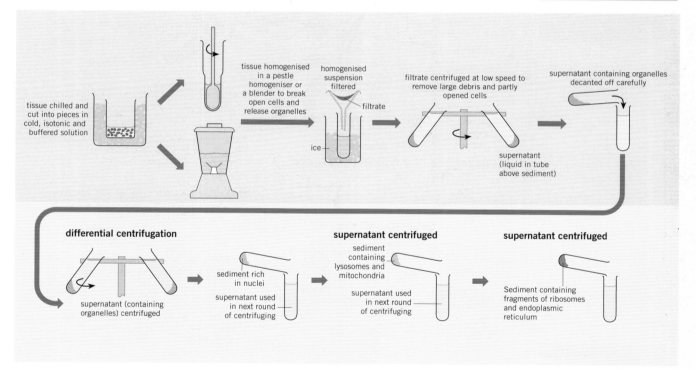

Cell fractionation and differential centrifugation

The ultrastructure of cells

Using an electron microscope, you can see the **ultrastructure** of the cell. This means that you can identify most of the organelles that the cell contains – far more than you can see with a light microscope.

Compare these animal and plant cell ultrastructure diagrams with those seen under the light microscope on page 10. What differences can you see?

GURU TIP

Questions about organelles and the ultrastructure of cells often appear on exam papers. Use the photomicrographs on the following pages to learn and test yourself on identifying cell organelles and their functions.

KEY SKILLS
C3.1b, IT3.3, WO3, LP3

Labels (animal cell): cytoplasm; centrioles - these are important when the cell divides; nuclear membrane; nucleus; nucleolus; smooth endoplasmic reticulum; chromatin; ribosomes; microfilaments; microtubles; outer cell membrane; secretory granules; intercellular space; Golgi body – secretes molecules into the secretory granules; nuclear pore; rough endoplasmic reticulum; mitochondrion

Ultrastructure of a typical animal cell

Try and learn how do these diagrams differ from those drawn using a high-quality light microscope, like those pictured on page 11.

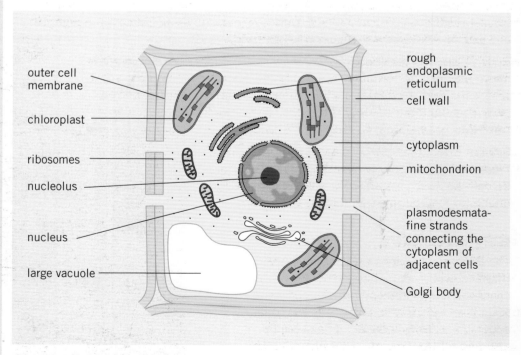

Labels (plant cell): outer cell membrane; chloroplast; ribosomes; nucleolus; nucleus; large vacuole; rough endoplasmic reticulum; cell wall; cytoplasm; mitochondrion; plasmodesmata- fine strands connecting the cytoplasm of adjacent cells; Golgi body

Diagram of the ultrastructure of a typical plant cell

The cell surface membrane

The **cell surface membrane** (or **plasma membrane**) is about 7nm thick and is made up of a highly organised layer system. This membrane controls the exchange of chemical materials between the cell and its environment. Later on, the cell surface membrane section looks at the structure of the membrane and how it performs its function.

Microvilli

Microvilli are extensions of the cell surface membrane, which increase the cell surface area. They are commonly found in cells that have a high absorption capacity, such as in kidney nephrons. The microvilli represent a **brush border** to the cell.

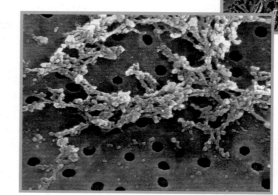

Coloured scanning electron micrograph of microvilli in an intestinal epithelial cell

Nuclear envelope

The nucleus is encased by a double membrane **nuclear envelope**. This membrane is perforated by large **nuclear pores**, through which chemical materials can be exchanged between the nucleus and the cytoplasm. The outer membrane joins with the **endoplasmic reticulum**. This is shown here, from above. Compare this to the cross section view, in the diagram on page 14.

Coloured scanning electron micrograph of nuclear pores and chromatin fibres

Endoplasmic reticulum

The endoplasmic reticulum (ER) is a system of membranes forming inter-connecting sacs or sheets, called **cisternae**. ER forms a transportation network that extends throughout the cell. There are two forms: rough endoplasmic reticulum (**rough ER**) and smooth endoplasmic reticulum (**smooth ER**).

Coloured scanning electron micrograph of rough ER

Rough ER has many **ribosomes** attached to its surface. The ribosomes are where proteins are manufactured inside the cell (you can find out about this in the DNA section). Rough ER transports newly manufactured proteins around the cell and is responsible for helping more complex, three-dimensional protein structures to form.

Smooth ER is where steroids and lipids form within the cell and is responsible for their transportation. Smooth ER forms a surface for enzymes systems to attach to. Find out more in the enzyme section.

Golgi apparatus

The **Golgi apparatus** is a stack of **cisternae** (flattened sacs) formed by membranes. It is constantly being formed and reformed. Small vesicles pinch off from one end of the Golgi apparatus as **Golgi vesicles**. Small vesicles, also bud off from the endoplasmic reticulum to join with the Golgi apparatus.

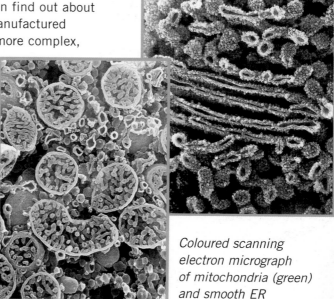

Coloured scanning electron micrograph of mitochondria (green) and smooth ER

KEY SKILLS C3.1b, IT3.3, W03, LP3

KEY SKILLS
C3.1b, IT3.3, WO3, LP3

The Golgi apparatus accepts the products of cell metabolism via the vesicles originating from the ER. If there are any misdirected molecules within the vesicles, they are moved back to the ER (again, inside vesicles). Product molecules are moved through the cisternae in a precisely defined sequence. As they move, the chemicals are often modified. Some of the Golgi vesicles that move away from the Golgi apparatus contain chemicals that will be excreted in the process of **exocytosis** (you can find out more about this, and about **endocytosis**, in the cell membrane section). Other Golgi vesicles move to other parts of the cell where their contents will be used in other processes.

Coloured scanning electron micrograph of Golgi apparatus

Lysosome

Lysosomes are special vesicles, bounded by a single membrane, and contain **hydrolytic** (digestive) enzymes. These may be used to digest materials taken into the cell (by endocytosis) or defective organelles. Lysosomes can also act as a 'self-destruct' mechanism for the cell, if their contents are released into the cytoplasm.

Peroxisome

Peroxisome is a member of a group of vesicles that contain oxidative enzymes, catalase, for example. These enzymes are important in delaying cell ageing.

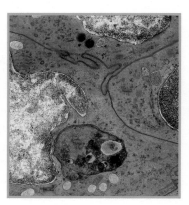

Coloured scanning electron micrograph of a secondary lysosome digesting a particle

Microfilament

Microfilaments consist of threads of the protein, **actin**. These bundle together beneath the cell surface membrane and help in the processes of endocytosis and exocytosis. Microfilaments also help some cells to move.

Microtubule

Microtubules are made up of hollow tubes of, **tubulin**. They form a cellular skeleton, to which other organelles can attach. Microtubules are used for movement by some cells and are also found in the structure of **cilia** and **flagella**.

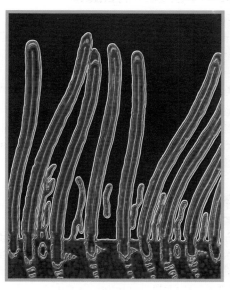

Computer graphic of a cross-section through cilia

Nucleus

The nucleus controls all the cell activities, since it contains the hereditary material, DNA. DNA carries the information necessary for the manufacture of proteins. We will look at the importance of this in the DNA section. DNA is loosely bound up with a protein called histone, to form **chromatin**. The nucleus contains one or more **nucleoli** in which **messenger RNA** (mRNA) and **transfer RNA** (tRNA) are manufactured.

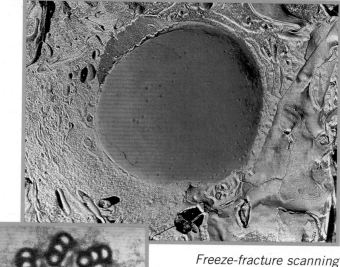

Freeze-fracture scanning electron micrograph of a nucleus

Centriole

Centrioles are only found in animal cells. They are organelles which appear in pairs (usually one pair per cell), held at right-angles to each other. They are made up from a number of microtubules. Centrioles are responsible for the organisation of the nuclear spindle during the process of **cell division**. You can find out about this process in the reproduction section.

Transmission electron micrograph of a centriole

Mitochondrion

A **mitochondrion** (plural: mitochondria) is surrounded by a double-membrane. The outer membrane is smooth but the inner membrane is folded inwards to make **cristae**. The space on the inside of the cristae membrane is known as the **matrix**. The folded cristae give the inner membrane a large surface area. The inner surfaces of the cristae are also covered in many small granules. These consist of enzymes that are used in the process of energy production in aerobic respiration. You can find out about this process in section about enzymes.

Specific plant organelles

Apart from centrioles, which are only found in animal cells, all the above organelles are found in both plant and animal cells. The organelles described below are only found in plant cells.

Cell wall

The cell wall consists of long **cellulose** molecules grouped into bundles called **microfibrils**, which are twisted into rope-like **macrofibrils**. The cell wall gives physical strength to a plant cell. Look into this in the cell membranes section.

Plasmodesmata

Plasmodesmata are minute strands of cytoplasm (including ER), which pass through pores in the cell wall and connect the contents (protoplasm) of adjacent cells. This allows for the movement of water and dissolved substances through the plant and helps cells to survive periods of drought.

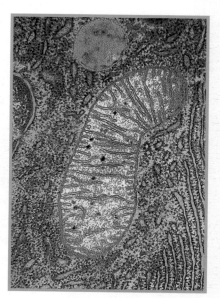

False-colour transmission electron micrograph of a mitochondrion

KEY SKILLS C3.1b, IT3.3, WO3, LP3

Middle lamella

The **middle lamella** contains gums and calcium pectate and form a 'biological glue'. This bonds adjacent cells together, giving extra, collective strength to a plant.

Vacuole

A **vacuole** can form up to 90% of the volume of plant cells. It is filled with **cell sap**, which is a solution of salts, sugars and organic acids. The cell sap helps to maintain the **turgor pressure** inside the cell (find out more in section 3.5). The vacuole is bounded by a membrane called the **tonoplast**.

Chloroplast

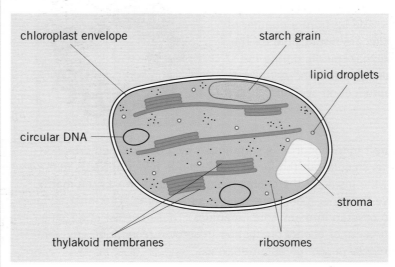

KEY SKILLS
C3.1b, IT3.3, W03, LP3

chloroplast envelope

starch grain

lipid droplets

circular DNA

stroma

thylakoid membranes

ribosomes

Detailed diagram of a chloroplast (magnification x 11 000)

Each **chloroplast** is surrounded by an envelope formed by two membranes. Inside the chloroplast, there is a fluid called **stroma**, which contains a series of flattened stacks of **thylakoids** (fluid-filled membranes). The stacks of thylakoids are called **grana**, which are also joined together by membranes. This arrangement of membranes allows chloroplasts to be very efficient at absorbing the energy contained within light during **photosynthesis**. You can find out more about chloroplasts and their role in photosynthesis in the enzymes section.

Prokaryotic and eukaryotic cells

Eukaryotic cells have a nucleus, containing their DNA, inside a nuclear envelope.

There are groups of organisms (bacteria and blue-green algae), which don't have DNA bound within a nuclear envelope. This much simpler type of cell is referred to as a **prokaryotic cell**. It is generally thought that eukaryotes evolved from prokaryotes at some dim, distant point back in the history of life on Earth.

Prokaryotes are important organisms in ecosystems. Many are involved in the decay of organism remains and the recycling of elements. Others can 'fix' atmospheric nitrogen and convert it into a suitable form for absorption into plants. Bacteria have also become an important tool in the process of **genetic engineering**.

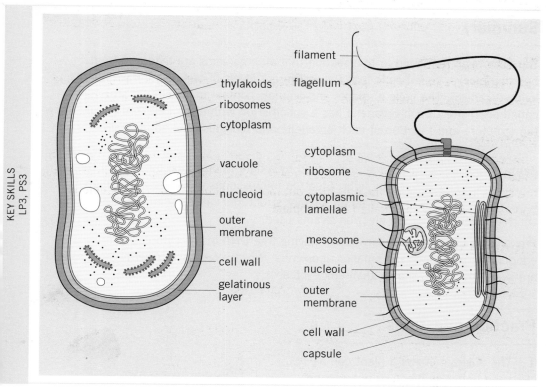

Structure of two generalised prokaryotic bacterial cells

thylakoids
ribosomes
cytoplasm
vacuole
nucleoid
outer membrane
cell wall
gelatinous layer

filament
flagellum

cytoplasm
ribosome
cytoplasmic lamellae
mesosome
nucleoid
outer membrane
cell wall
capsule

GURU TV
Find out how AS Biology expert, Mary Jones, defines the difference between a prokaryotic and a eukaryotic cell.

Feature	Prokaryotic cell	Eukaryotic plant cell	Eukaryotic animal cell
diameter	0.5–5μm	up to 40μm	
organisation	single cells	generally part of a tissue	
cell wall	made of a polysaccharide (but not cellulose)	made of cellulose	not present
nucleus	not present	present, surrounded by an envelope of 2 membranes	
DNA	single, circular thread	several linear chromosomes associated with protein formation	
ribosomes	small ribosomes, 18nm in diameter	larger ribosomes, 22nm in diameter	
ER	not present	present	
mitochondria	not present	present	
chloroplasts	not present	present	not present
phospholipid plasma membrane	present	present	

Comparison of pro and eukaryotic cells

Summary

You have been reminded in this section that all organisms are made up from building blocks, called cells that show different levels of organisation within any living organism. The cells of plants and animals contain many similar structures called organelles. Each organelle has a specific job within the cell. You should know the structure and function of each organelle and recognise the additional parts only found in plant cells.

Some organelles are too small to be seen by a light microscope because of the limited resolution offered by light. Electron microscopes provide a much higher resolution, and consequently much smaller structures can be seen, because of the shorter wavelength of the electron beams used.

Prokaryotes differ from eukaryotes in having free DNA in the cytoplasm, no edoplasmic reticulum and smaller ribosomes. Prokaryotes have fewer organelles and are generally much smaller than eukaryotes.

Practice questions

1   The diagram shows a plant cell as seen under an electron microscope.

    a Identify structures A to H.

    b Which structures would be absent from an animal cell?

    c An electron microscope has a much higher resolving power than a light microscope. Explain what the word 'resolution' means.

2   List four features that prokaryotic and eukaryotic plant cells have in common. For each, briefly explain its function within the cell.

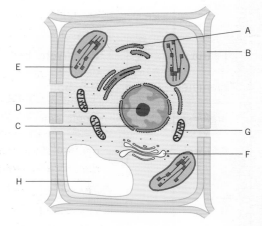

3   Describe how you would isolate the mitochondria from a piece of liver tissue.

Key skills

If you can describe and interpret drawings and photographs of typical plant and animal cells as seen under the light and electron microscopes, you have opportunities to demonstrate the C3.1b, IT3.3, WO3 (all), LP3 (all) key skills. Also, if you can draw clear, labelled diagrams of cells and tissues and also calculate the magnification and size of structures seen, you will be able to show all the N3 and LP3 skills. Go through the checklist in the introduction of this book to see what you need to do.

Biological molecules

In this section you will learn about:

☞ how carbohydrates, proteins, lipids and polynucleotides are built from smaller building blocks that are repeatedly joined together

☞ relating the structure of biological molecules to their functions

☞ how different types of chemical bond give different properties to biological molecules

☞ the role and importance of water and any dissolved inorganic mineral salts to living organisms.

This section looks at the structure, properties and functions of important biological molecules found in living organisms: carbohydrates; proteins; nucleotides and lipids (the first three are referred to as **macromolecules**). It looks at how these are built up from smaller units (**monomers**), into complex organic molecules (**polymers**), which have huge biological implications. You will also see how and why the properties of water and dissolved minerals are so significant to living organisms.

An organic compound is one that contains carbon and hydrogen.

The four most common elements in living organisms are hydrogen, carbon, oxygen and nitrogen. Together, they account for more than 99% of atoms found in living organisms. Carbon in particular is important, as its atoms tend to bond together to form long chains or ring structures. These, in turn, tend to form a 'molecular skeleton' to which other atoms can bond.

Make sure that you have a really good understanding of biological molecules. This relates to all areas of your syllabus. At a basic level, you will have come across many of the important ideas about biological molecules before, at KS4.

Remember: try to revise what you already know about atoms, molecules and chemical bonding; carbohydrates, proteins, lipids (fats and oils); ATP and DNA, and build on this.

Students often find biochemistry difficult, so don't panic if you find it confusing at first. Work steadily at understanding each type of biological molecule in turn, making sure you understand that they are built up from simpler sub-units.

Chemicals of life: carbohydrates

A carbohydrate is part of a group of related substances, which just contain carbon (C), hydrogen (H) and oxygen (O).

GURU TIP
Carbohydrate names generally end with **-ose**.

Carbohydrates can be divided into three or more categories depending upon their level of complexity: **monosaccharides**; **disaccharides** and **polysaccharides**, with monosaccharides being the simplest. The table below shows the more common carbohydrates, their properties and how they can be converted to different forms.

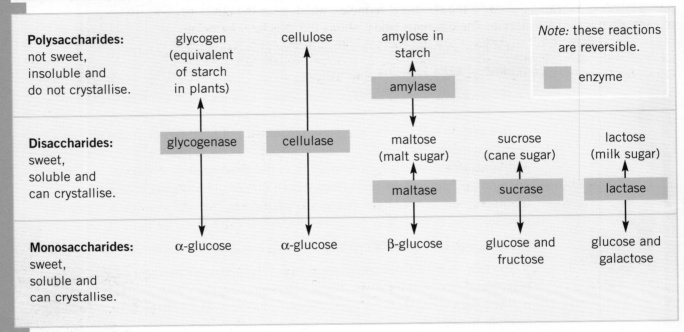

Polysaccharides: not sweet, insoluble and do not crystallise.	glycogen (equivalent of starch in plants)	cellulose	amylose in starch	*Note:* these reactions are reversible.	enzyme
			amylase		
Disaccharides: sweet, soluble and can crystallise.	glycogenase	cellulase	maltose (malt sugar)	sucrose (cane sugar)	lactose (milk sugar)
			maltase	sucrase	lactase
Monosaccharides: sweet, soluble and can crystallise.	α-glucose	α-glucose	β-glucose	glucose and fructose	glucose and galactose

Summary of the common carbohydrates and their functions

Monosaccharides

Monosaccharide (and disaccharide) carbohydrates are sugars and all have the general formula of $(CH_2O)_n$, where n represents the number of times the CH_2O unit is repeated (and the number of carbon atoms within the molecule). This formula shows the relative proportions of the three elements within the carbohydrate molecule.

KS4 chemistry
Different atoms have different valencies (number of bonds):
carbon – 4
nitrogen – 3
oxygen – 2
hydrogen – 1.

Monosaccharide carbohydrates can be further classified according to how many carbon atoms they contain:

3 Carbon atoms = triose sugars, for example glyceraldehyde

5 Carbon atoms = pentose sugars, for example fructose, ribose, deoxyribose

6 Carbon atoms = hexose sugars, for example glucose, galactose

Monosaccharides have two main functions in living organisms:

1 They are used as a source of energy in respiration. Their molecules contain many carbon-hydrogen bonds that can be broken to release energy.

2 They are used as building blocks for larger molecules. For example, ribose is used to make ribonucleic acid (RNA) and adenosine triphosphate (ATP); deoxyribose is used to make DNA. You can find out more in the enzyme and DNA sections.

GURU TIP
For simplicity, C and H atoms are often left out of structural formulae, don't let this confuse you!

Molecular and structural formulae

The **molecular formula** for glucose ($C_6H_{12}O_6$) for example, doesn't give any idea of how the atoms are arranged within the molecule. A much more helpful method of presentation is a diagram, showing the **structural formula**, which plots the positions of each atom within a molecule. The diagrams below, shows the structural formulae of glucose. Glucose has two forms: alpha (α) and beta (β). It exists either as a straight chain molecule or as a ring structure, and these structures are reversible. Generally, the ring structure is more stable when it is in solution.

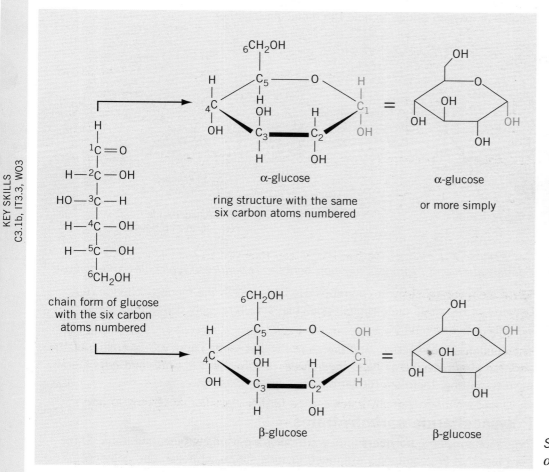

chain form of glucose with the six carbon atoms numbered

α-glucose
ring structure with the same six carbon atoms numbered

α-glucose
or more simply

β-glucose

β-glucose

Structural formulae of α– and β-glucose

GURU WEBSITE
There's a great section on this reaction on the AS Guru™ Biology website. You can drag molecules into the reaction and see what happens.

The number of carbon atoms within pentose and hexose molecules is enough for the molecule to close up and form a much more stable ring structure.

You will find it helpful to number the carbon atoms within a ring structure. The convention is to work round in a clockwise direction, starting on the right.

Biological molecules

KEY SKILLS
C3.1b, IT3.3, WO3

Chemicals of life: carbohydrates

To help you understand the three-dimensional nature of molecules, lines representing bonds between carbon atoms are drawn thickly at the front of the molecule.

α- and β-glucose have the same molecular formula, however their structural formulae are different in one important way. Notice that the **hydroxyl** group in carbon position 1 of the α-glucose molecule has been rotated by 180° in position 1 of the β-glucose molecule, (α- and β-glucose are **isomers** of each other). This gives them different chemical properties.

Indicating the difference between α- and β-glucose

Disaccharide carbohydrates and the glycosidic bond

Two monosaccharide carbohydrate molecules may join together to form a larger, single disaccharide molecule (see the diagram below). This requires a process called **condensation**, because it results in the removal of one molecule of water. The resulting link between the two monosaccharides, is known as a **glycosidic bond** and creates an **oxygen bridge** to hold the molecule together.

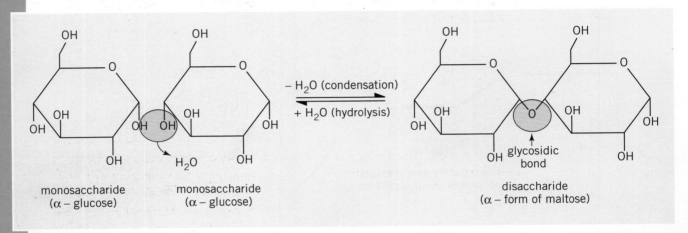

monosaccharide (α – glucose)   monosaccharide (α – glucose)   disaccharide (α – form of maltose)

− H₂O (condensation)
+ H₂O (hydrolysis)

glycosidic bond

Condensation and hydrolysis reaction between two glucose molecules and maltose

There are a variety of hydroxyl groups available within the monosaccharide, but because of the three-dimensional molecular structure, only bonding at positions 1,4 and 1,6 are common.

The condensation reaction, and therefore the glycosidic bond, are reversible by the addition of a molecule of water (**hydrolysis** – *hydro* refers to water and *lysis* literally means splitting, therefore the term means *water-splitting*).

Polysaccharide carbohydrates

Polysaccharides are **polymers** – they are made up of many repeating units. Three polymers that are commonly examined include starch, glycogen and cellulose.

GURU TIP

Examiners like to ask you to explain how polysaccharide structure relates to function. Remember that they are insoluble in water and therefore have no osmotic effect on water potential and are unable to diffuse out of the cell.

Polysaccharides	Structure	Function
starch	Made of two polymers of α-glucose: Amylose: a chain of α-glucose molecules joined by 1,4 glycosidic bonds. The position of the `O` in the glycosidic bond and hydrogen bonding (overleaf), causes the long molecule to spiral into a helix.	Main storage polysaccharide in plants. The helix is compact, making it excellent for storage.
	Amylopectin: a chain of α-glucose molecules joined by 1,4 glycosidic bonds but every 25 glucose units a 1,6 glycosidic bond takes place. This causes the molecule to branch.	The side branches can be hydrolysed quickly, giving rapid release of glucose for energy production in the process of respiration (page 73)
glycogen	This is similar in structure to amylopectin except that there are many more 1,6 glycosidic side branches.	A very compact molecule, which is found in the mammalian liver and fungal cells. The increased branching structure allows for quick hydrolysis to release glucose. This means that energy can be released more quickly in animals, through the process for respiration (page 73)
cellulose	This is made of long, unbranched chains of β-glucose, joined by 1,4 glycosidic bonds. Individual chains are linked to each other by hydrogen bonds (overleaf). These are formed into long, strong fibres, called microfibrils. Notice the alternating `O` `O` `O` position of the glycosidic bond and compare it to amylose	The structural polysaccharide in plants. Cellulose cell walls provide strength to plant cells. The hydrogen bonding between cellulose chains prevent water form entering the molecule. This makes it resistant to hydrolysis by enzymes.

Summary of starch (amylose and amylopectin), glycogen and cellulose

Chemicals of life: proteins

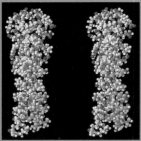

Molecular graphic of keratin – found in skin and hair

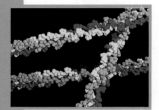

Molecular graphic of collagen, found in connective tissues

Proteins form more than 50% of the dry mass of cells and are therefore very important constituents of living organisms. Here are some of the important jobs they do:

- enzymes help control all the chemical reactions taking place within the organism
- they play an important role in allowing molecules to move through cell membranes
- they are an important part of transport systems, for example the blood pigment, **haemoglobin**, is used to transport oxygen in the blood
- many hormones carried in blood or plant transport systems are based on proteins, which trigger responses in other parts of the organism, for example the control of blood glucose levels by insulin
- hair, nails and skin contain the protein **keratin**
- antibodies neutralise invading micro-organisms, which may cause harm, and are made from different types of protein
- proteins help to maintain the pH of blood by buffering the solution
- **actin** and **myosin**, help muscle tissues to contract
- many tissues are strengthened by collagen, for example bone and artery walls.

The building blocks of proteins: amino acids

Proteins are built up from smaller sub-units called **amino acids**, about 20 of which occur naturally. These can join, end-to-end, to form long chains. The order and number of the amino acids in the chain determine the properties of the protein. There are large numbers of possible types of protein – you will be pleased to hear that you don't have to learn many of them.

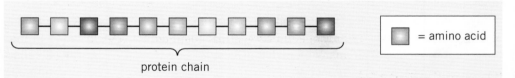

protein chain

☐ = amino acid

Simple diagram of protein chains

The different bonds that are important in all biological molecules

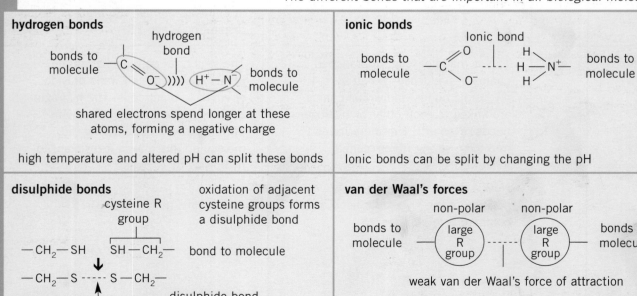

hydrogen bonds

bonds to molecule

hydrogen bond

bonds to molecule

shared electrons spend longer at these atoms, forming a negative charge

high temperature and altered pH can split these bonds

ionic bonds

Ionic bond

bonds to molecule

bonds to molecule

Ionic bonds can be split by changing the pH

disulphide bonds

cysteine R group

oxidation of adjacent cysteine groups forms a disulphide bond

bond to molecule

—CH₂—SH   SH—CH₂—

—CH₂—S ---- S—CH₂—

disulphide bond

disulphide bonds can be split by reducing agents

van der Waal's forces

non-polar   non-polar

bonds to molecule

large R group

large R group

bonds to molecule

weak van der Waal's force of attraction

these forces can be split by a rise in temperature

Amino acids all have the same basic type of structure, differing only in the atoms bonded to the central carbon atom. This is known as the **R-group** (or side chain) and can be as simple as a single hydrogen atom (glycine). Sometimes, the most important function of the R-group is to determine how the 3-dimensional structure of a protein is held together and therefore deciding its function.

R group or side chain varies in different amino acids.

amine group can pick up a H$^+$ ion from surroundings because it is a base

carboxylic acid group loses a H$^+$ ion to surroundings

glycine molecule

alanine molecule

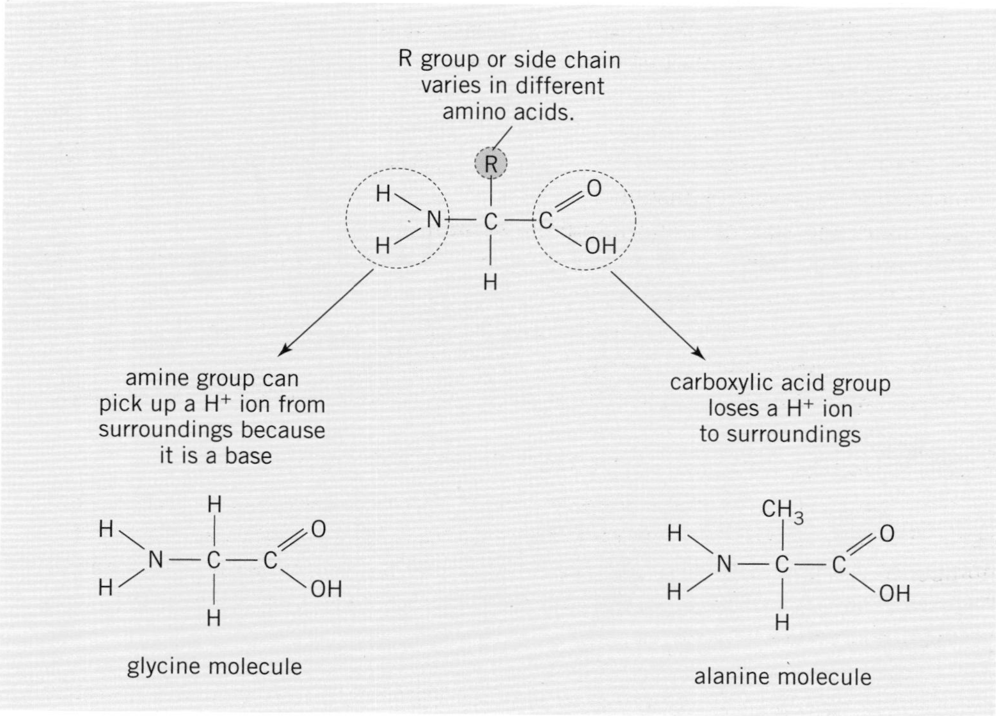

Structure of a typical amino acid and examples of common amino acids

The peptide bond

Amino acids are held together by **peptide bonds**. Like the glycosidic bond between carbohydrate sub-units, peptide bonds are formed by a condensation reaction. The carboxylic acid group of one amino acid lines up with the amine group of a second amino acid. One loses a -OH (hydroxyl) group from its carboxylic acid group, while the other loses a hydrogen atom from its amine group. This leaves a carbon atom to combine with a nitrogen atom and it is this peptide bond that holds the two amino acids together as a **dipeptide**. The process can continue with further amino acids bonding to either end of the dipeptide, to form a **polypeptide** chain (protein).

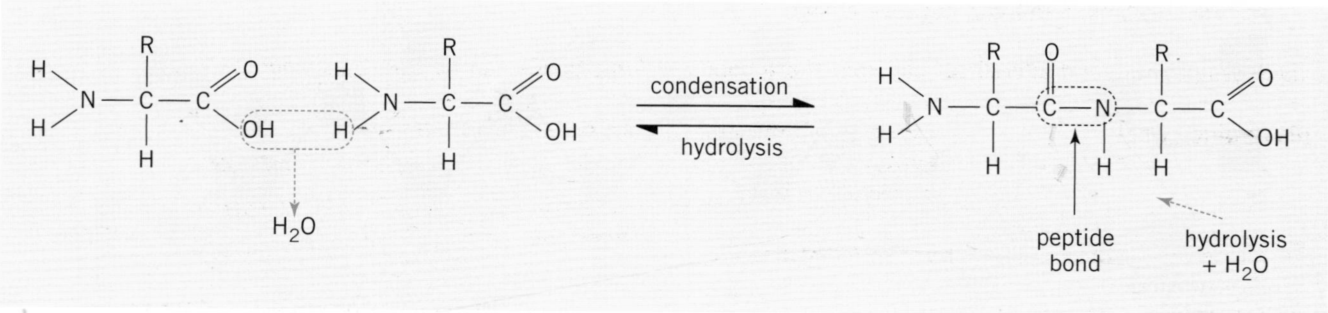

condensation

hydrolysis

H_2O

peptide bond

hydrolysis + H_2O

Dipeptide formation and the peptide bond

Biological molecules

Chemicals of life: proteins

GURU TIP

Compare hydrogen bonding in proteins with hydrogen bonding in carbohydrates (page 27). You should see some similarities.

Primary protein structure

A polypeptide molecule can contain from a few, to many thousands of amino acids. In each case, a chain is formed with a particular number and order of amino acids. This sequence is known as the **primary structure**. There are an enormous number of possible primary structures, each having different physical and chemical properties.

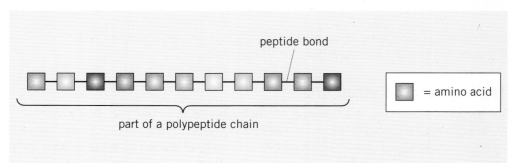

Primary structure of proteins

Secondary protein structure

The amino acids in a polypeptide chain (or between chains) interact with each other using various types of bonds. **Hydrogen bonding** is important in forming the **secondary structure** of proteins. Here, the -CO (carboxyl) group of one amino acid (the oxygen has a slight negative charge, due to electron sharing with the carbon) is attracted to the -NH (amine) group of another amino acid (the hydrogen has a slight positive charge due to electron sharing with the nitrogen). Alternatively, hydrogen bonding can take place between -CO and -OH (hydroxyl) groups (where the hydrogen has a slight, positive charge). Hydrogen bonding is dependent upon the nature of the amino acid R-group.

Two common secondary protein structures include the α-**helix** and β-**pleated sheet**.

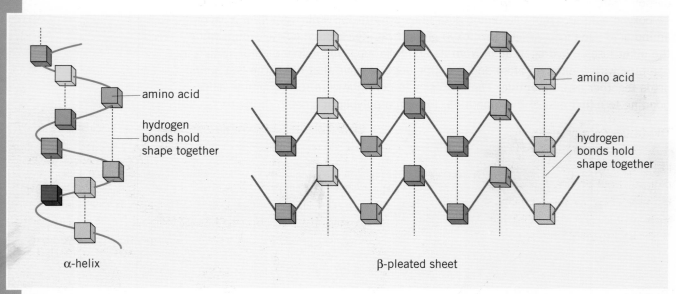

Polypeptide chains form an α-helix and β-pleated sheet

In the α-helix, the -CO group of one amino acid forms a hydrogen bond with the -NH group of another amino acid, four places ahead. In the β-pleated sheet, hydrogen bonds are formed between adjacent amino acids in different polypeptide chains, forming a layered structure, which looks a bit like a corrugated roof. This is an example of inter-polypeptide bonding.

> **Remember:** hydrogen bonds are weak and can be broken by excessive heat and changes in pH.

Some polypeptide chains show no secondary structure, depending on which R-groups are present and their position in the chain.

Tertiary protein structure

Many proteins show yet another level of complexity, where additional bonds form between R-groups of amino acids in different parts of the polypeptide chain. These forces give the protein a very precise three-dimensional shape, which is specific to a particular protein. This is known as the **tertiary structure**. Because of this, a tertiary structure may show tighter coiling, bends and sections where the polypeptide chain is more stretched out.

The main types of bonding leading to the different structures of protein are **hydrogen**, **ionic**, **disulphide** bonds and **van der Waal's** forces. You will need to learn how these can be affected by conditions such as temperature and pH changes (page 27).

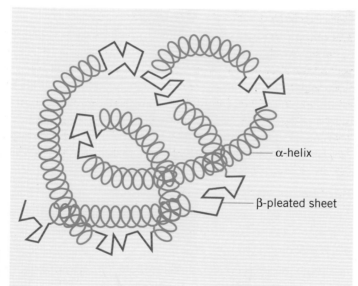

Three-dimensional structure of myoglobin

α-helix

β-pleated sheet

The tertiary structure shows a precise three-dimensional structure, which relates to just one polypeptide chain. Its shape is determined by the amino acid sequence on the chain and the R-groups they contain.

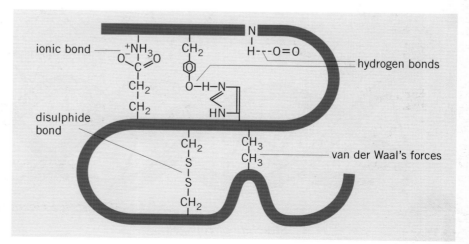

ionic bond

disulphide bond

hydrogen bonds

van der Waal's forces

Bonding between polypeptide chains

Chemicals of life: proteins

Quaternary protein structure

Some proteins consist of more than one polypeptide chain that are held together in a precise three-dimensional structure as tertiary proteins. They are held together by the same types of forces. These proteins show a **quaternary structure**.

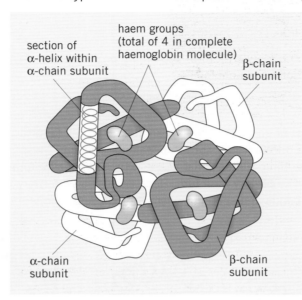

section of α-helix within α-chain subunit

haem groups (total of 4 in complete haemoglobin molecule)

β-chain subunit

α-chain subunit

β-chain subunit

Quarternary structure of haemoglobin

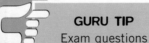

GURU TIP

Exam questions often test your understanding of protein structure by using the example of haemoglobin.

Haemoglobin consists of four polypeptide chains (two identical α-chains and two β–chains). It also contains a **prosthetic group** (one that is not made of amino acids) called **haem**, which is made of four iron ions (Fe^{2+}). Haem is responsible for the red colour of haemoglobin and is what transports oxygen (four molecules of oxygen for each molecule of haemoglobin) in animal circulatory systems.

Globular and fibrous proteins

There are many categories of proteins but they fall into two groups: fibrous and globular. Look at the proteins on page 27. Decide whether they are fibrous or globular.

In **fibrous proteins**, molecules form long chains or fibres (primary, secondary, tertiary and quaternary structures). Their fibrous nature tends to make the molecules insoluble in water and this makes them useful for structure and support.

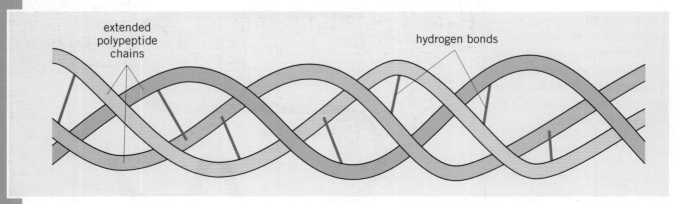

extended polypeptide chains

hydrogen bonds

Structure of collagen

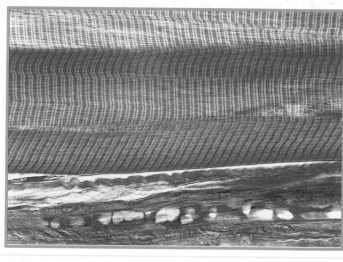

Skeletal muscle contains actin and myosin (striping) and collagen (blue)

KEY SKILLS
C3.1b, IT3.3, W03

Collagen, for example, is found in the skin, teeth, skeleton, tendons and blood vessel walls. The fibres form a triple-helix of polypeptide chains held together by hydrogen bonds, making them extremely strong.

In **globular proteins**, the molecules form a spherical mass with a specific three-dimensional structure (tertiary and quaternary structures). They curl up, so that any **hydrophillic** (water-loving) side chains are arranged on the outside of the molecule and any **hydrophobic** (water-disliking) side chains are arranged on the inside of the molecule. This makes the whole molecule soluble in water and ideal for transport. Haemoglobin and myoglobin molecules are examples of globular proteins.

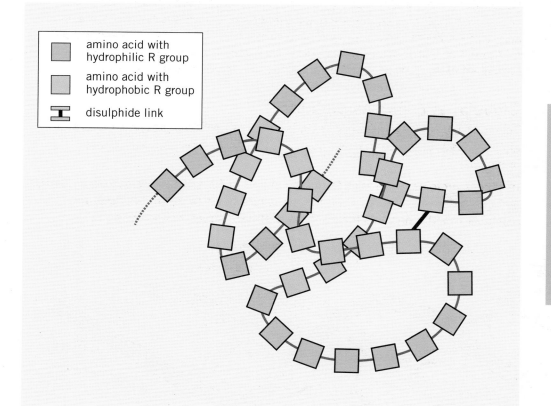

amino acid with hydrophilic R group

amino acid with hydrophobic R group

disulphide link

Schematic diagram of a section through a globular protein

Chemicals of life: lipids

Lipids, like carbohydrates, are made up of carbon, hydrogen and oxygen atoms, only in different proportions. Fats, oils and waxes are forms of lipids that all have different melting points.

Properties and functions of lipids

- Lipids are a great energy source. Mass for mass, lipids release about twice as much energy as carbohydrates (for example, if a mass of a lipid releases 38 kJ/g, the same mass of carbohydrates would release 17 kJ/g).
- They consist of compact molecules that are insoluble in water. This makes them ideal for storage.
- They are poor conductors of heat, which makes them ideal for maintaining body temperature by storing lipids under the skin.
- Delicate organs are supported by lipid layers.
- Oils and waxes help to prevent water evaporation through skin and plant leaves.
- Phospholipids are a major component of cell membranes.
- Lipids can form a layer of electrical insulation around nerve cells (neurones).
- When oxidised in respiration, water and carbon dioxide are released. This water may be of real importance to some organisms, allowing them to survive in very dry habitats.

Triglycerides

Most fats and oils are **triglyceride** lipids, formed by **esterification** (combining **fatty acids** and **glycerol** in a **condensation** reaction). If three fatty acid molecules join to one glycerol molecule the resulting lipid is a triglyceride.

Lipid formation is a result of a condensation reaction. Compare this with carbohydrate and protein formation in previous sections.

Fatty acids

There are seventy different types of fatty acids, each with the same basic structure. At one end of the molecule, there is a **carboxyl** group (-COOH). This is followed by a backbone of 15–17 carbon atoms with hydrogen atoms attached. These long 'tails' formed by the carbon chain backbone are hydrophobic and therefore don't dissolve in water. The tails are **non-polar** or **neutral** (there is an even distribution of electrical charge within the molecule).

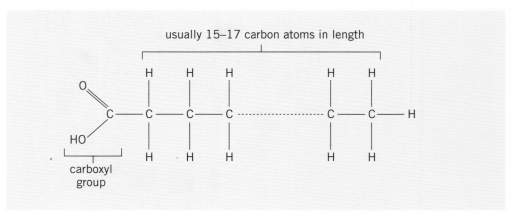

Typical structure of a fatty acid

If each carbon atom in the backbone has all the available bonds satisfied with hydrogen atoms, the fatty acid is said to be **saturated** (saturated with hydrogen) and will have the general formula $C_nH_{2n}O_2$. If some adjacent carbon atoms in the backbone form a double bond (C = C), then the fatty acid is known as **unsaturated** since it doesn't hold all the hydrogen that it could. Unsaturated lipids are generally solid at room temperature and tend to come from animal fats, whereas saturated lipids tend to come from plants, in the form of oils, which are liquid at room temperature, for example sunflower oil.

Unsaturated fatty acid molecules tend to have a 'kink' in the carbon atom backbone, whereas saturated fatty acid molecules tend to be much straighter.

If only one pair of carbon atoms forms this double bond, then the fatty acid is called **monounsaturated**, if there are two pairs of carbon atoms with double bonds, then it is called **polyunsaturated**.

Because the fatty acid chains can be very long, a shorthand way of representing them is used. The carboxyl group is shown and the rest of the molecule is drawn as a zig zag line:

/\/\/\/\/\/COOH or

/\/\/\/ = \/\/COOH

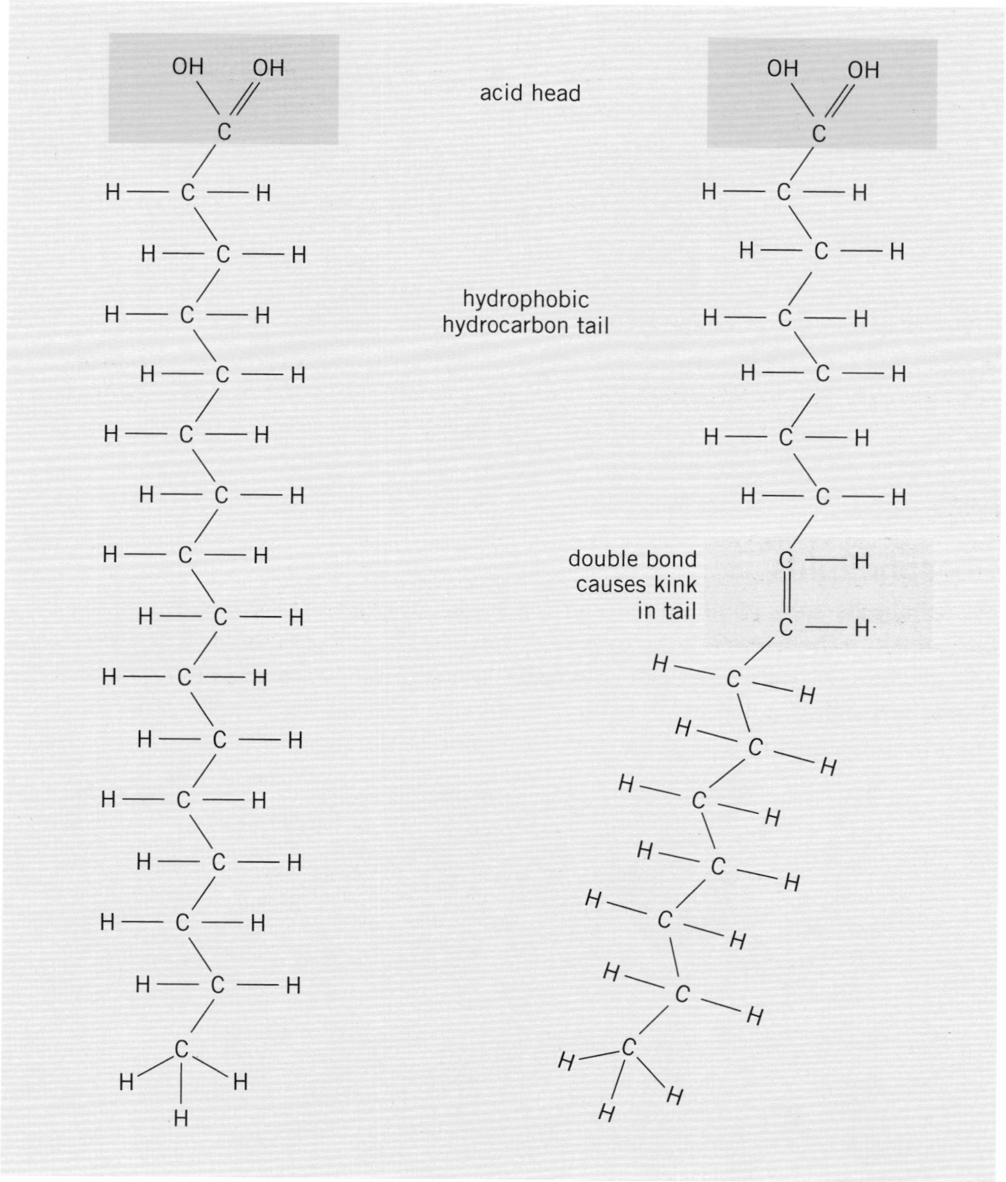

Saturated and unsaturated fatty acid molecules

Chemicals of life: lipids

Glycerol

Glycerol ($C_3H_8O_3$) is a form of alcohol. Notice the three hydroxyl (-OH) groups. An **ester** bond is formed between one of the glycerol's hydroxyl groups and the fatty acid's carboxyl group. A water molecule is released – another condensation reaction.

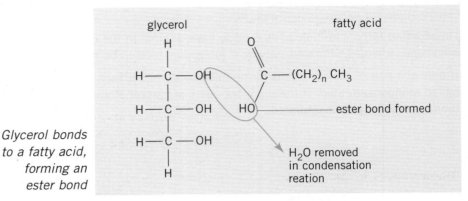

glycerol   fatty acid

Glycerol bonds to a fatty acid, forming an ester bond

ester bond formed

H_2O removed in condensation reation

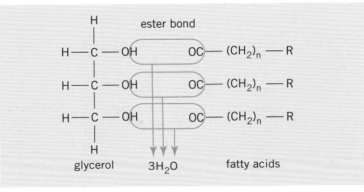

ester bond

glycerol $3H_2O$ fatty acids

Glycerol bonds to three fatty acids, forming a triglyceride by esterification

If only one fatty acid combines with the glycerol, it's called a **monoglyceride**. If two fatty acids combine, it is called a **diglyceride** and if three fatty acids combine, a **triglyceride**.

Phospholipids

Phospholipids are a common and important part of cell membranes. One of the fatty acids in the triglyceride molecule is replaced by **phosphoric acid** (H_3PO_4). The resulting phospholipid molecule has a charged (polar) head, which makes it hydrophilic and two non-polar, hydrophobic tails. You will discover how this is important to the structure of the cell membrane in the next section.

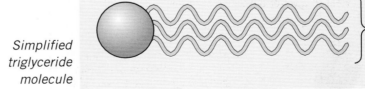

Simplified triglyceride molecule

3 non-polar hydrophobic fatty acid tails

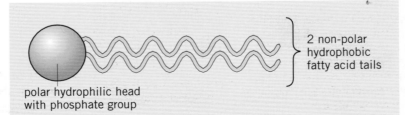

polar hydrophilic head with phosphate group

2 non-polar hydrophobic fatty acid tails

Simplified phospholipid molecule

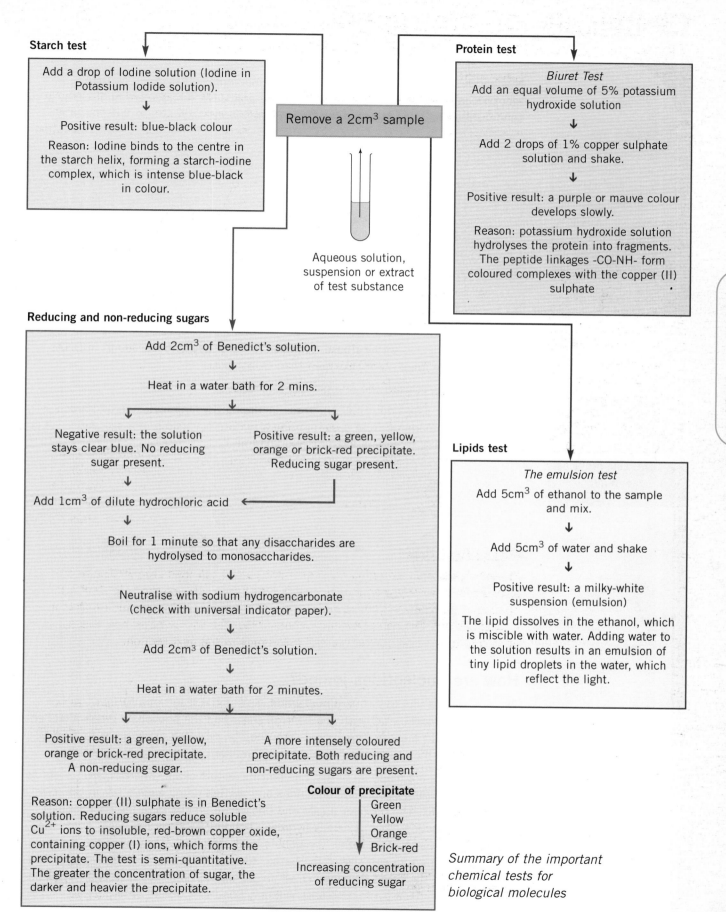

Starch test

Add a drop of Iodine solution (Iodine in Potassium Iodide solution).

↓

Positive result: blue-black colour

Reason: Iodine binds to the centre in the starch helix, forming a starch-iodine complex, which is intense blue-black in colour.

Remove a 2cm³ sample

Aqueous solution, suspension or extract of test substance

Protein test

Biuret Test
Add an equal volume of 5% potassium hydroxide solution

↓

Add 2 drops of 1% copper sulphate solution and shake.

↓

Positive result: a purple or mauve colour develops slowly.

Reason: potassium hydroxide solution hydrolyses the protein into fragments. The peptide linkages -CO-NH- form coloured complexes with the copper (II) sulphate

Reducing and non-reducing sugars

Add 2cm³ of Benedict's solution.

↓

Heat in a water bath for 2 mins.

Negative result: the solution stays clear blue. No reducing sugar present.

Positive result: a green, yellow, orange or brick-red precipitate. Reducing sugar present.

Add 1cm³ of dilute hydrochloric acid

↓

Boil for 1 minute so that any disaccharides are hydrolysed to monosaccharides.

↓

Neutralise with sodium hydrogencarbonate (check with universal indicator paper).

↓

Add 2cm³ of Benedict's solution.

↓

Heat in a water bath for 2 minutes.

Positive result: a green, yellow, orange or brick-red precipitate. A non-reducing sugar.

A more intensely coloured precipitate. Both reducing and non-reducing sugars are present.

Reason: copper (II) sulphate is in Benedict's solution. Reducing sugars reduce soluble Cu^{2+} ions to insoluble, red-brown copper oxide, containing copper (I) ions, which forms the precipitate. The test is semi-quantitative. The greater the concentration of sugar, the darker and heavier the precipitate.

Colour of precipitate
Green
Yellow
Orange
Brick-red

Increasing concentration of reducing sugar

Lipids test

The emulsion test
Add 5cm³ of ethanol to the sample and mix.

↓

Add 5cm³ of water and shake

↓

Positive result: a milky-white suspension (emulsion)

The lipid dissolves in the ethanol, which is miscible with water. Adding water to the solution results in an emulsion of tiny lipid droplets in the water, which reflect the light.

Summary of the important chemical tests for biological molecules

Biological molecules

Chemicals of life: nucleotides

Nucleotides are the basic building blocks of **macromolecules**, such as **DNA** (**deoxyribonucleic acid**) and **RNA** (**ribonucleic acid**). Nucleotides are built up by a **condensation** reaction between the three components.

GURU TIP
Nucleotides are made up of three components: a **pentose sugar**, a **base** containing nitrogen and a **phosphate** group.

Structure of ribose and deoxyribose sugars

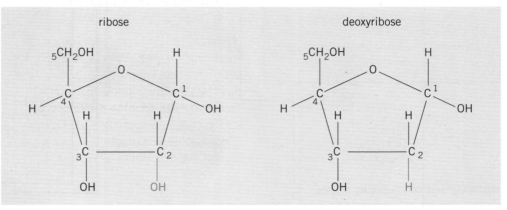

The pentose sugar is either **ribose**, found in RNA, or **deoxyribose**, found in DNA. Ribose has one more atom of oxygen than deoxyribose.

There are two groups of the nitrogen-containing bases:

GURU TIP
Uracil is only found in **RNA**, and **thymine** in **DNA**. One replaces the other.

- **pyrimidines** – six-sided ring structures including **thymine**, **cytosine** and **uracil**
- **purines** are made up from a double-ring structure consisting of a six- and five-sided ring, including **adenine** and **guanine**.

Simplified diagram of nucleotide formation

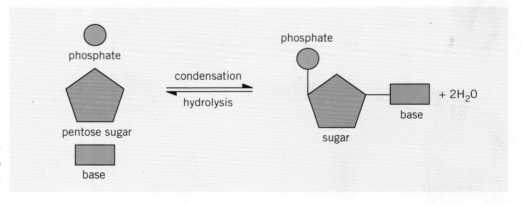

How are nucleotides joined?

Two nucleotides join to form a **dinucleotide**. This is the result of a further **condensation** reaction between one pentose sugar and the phosphate group attached to another pentose sugar. This process can then be repeated many times to form a **polynucleotide** molecule, which could contain millions of individual nucleotides.

RNA molecules are made up of just one polynucleotide chain, whereas DNA exists as two chains, running in opposite directions and twisted round each other in the form of a **double helix**. The pentose sugars and phosphate groups form the backbone of each polynucleotide chain, whereas **hydrogen bonding** between the bases holds the individual chains together within the double helix structure. Because of the three-dimensional structure and size of the bases, **adenine** (**A**) will only form hydrogen bonds between itself and **thymine** (**T**), while **guanine** (**G**) will only bond with **cytosine** (**C**). In other words, a pyrimidine base will only bond to a purine base.

KEY SKILLS
PS3, LP3

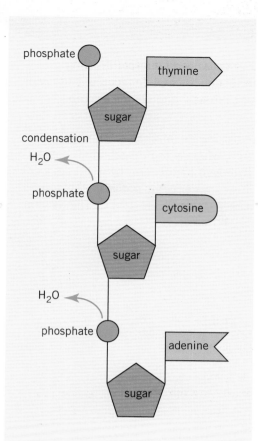

The sequence of bases along the polynucleotide chain and the **complimentary base pairing** forms the basis for the genetic code. This code allows a cell to manufacture specific proteins.

You will come across other biological molecules in your AS course which are also based on the nucleotide structure: **ATP** (**adenosine triphosphate**), **NAD** (**nicotinamide adenine dinucleotide**) and **NADP** (**nicotinamide adenine dinucleotide phosphate**).

Formation of a polynucleotide by a condensation reaction

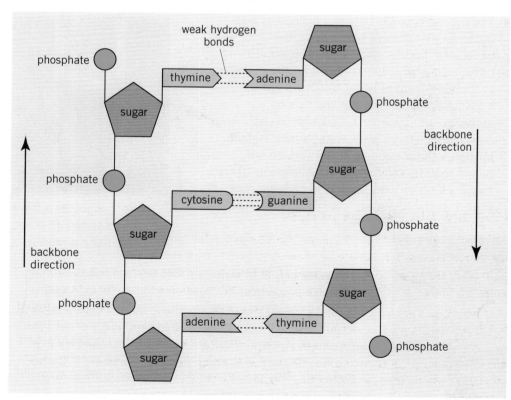

Base bonding between pyrimidines and purines

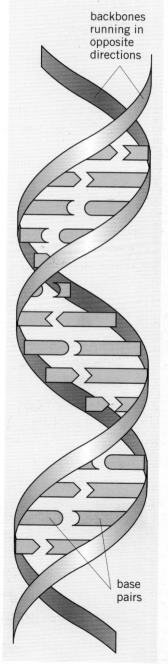

Double helix structure

Chemicals of life: water and minerals

Water (H_2O) is the most abundant chemical in cells, and therefore in living organisms (humans are about 60% water). Although a simple molecule, and easily taken for granted, water has the most amazing physical and chemical properties and supports all life on planet Earth.

GURU TIP

Water is a favourite area for essay-style exam questions, because it displays so many important properties. Take care not to waffle – be specific and keep to the point.

KEY SKILLS
C3.1b, IT3.3, WO3

- Water is a **polar** molecule with an uneven distribution of charge. The two hydrogen atoms have a small positive charge, whereas the oxygen atom has a small negative charge. As a result, weak **hydrogen bonds** form between adjacent water molecules. These hydrogen bonds makes water comparatively stable and allow it to stay in liquid form over the broad range of 0°C to 100°C. This is vital for living organisms.

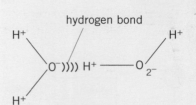

water molecule

weak hydrogen bond forming between two water molecules

Water is a polar molecule and forms hydrogen bonds with other polar molecules

- The hydrogen bonds between molecules give water strong **cohesive** properties – the molecules tend to 'stick' to each other. This is the reason why a siphon works, and is exploited in the movement of water through plant transport systems (see the sections on exchange and transport). Cohesive forces between molecules gives water a high **surface tension** – the molecules at the surface are pulled towards the body of water beneath. This forms a kind of 'skin', which some organisms exploit as their habitat.

Water molecules adhere strongly to glass, forming a meniscus

Water strider using surface water tension

- The polar nature of water also means that it has strong **adhesive** properties: its molecules like to 'stick' to other surfaces containing polar molecules. This is demonstrated by the meniscus at the top of a column of water, or the way water is drawn up in a capillary tube (capillary action or capillarity), or is drawn through chromatography paper (overleaf). This property also helps water to move through plant transport systems.

Polar qualities create adhesive properties in water

The hydrogen bonding between molecules is also responsible for the important thermal properties of water.

- Water has a high **specific heat capacity** (it requires a great deal of energy to raise its temperature or a lot of energy needs to be lost in order for the temperature to fall). As a result, the high water content of cells helps them to maintain their temperature. Water is also used to transport heat around an organism.
- Water has a high **latent heat of vaporisation** so that when water molecules evaporate from solution, a great deal of energy is released with them. This is used in living organisms as a cooling mechanism in sweating, for example.
- Ice is less dense than water and therefore floats. As the temperature of the water decreases, the molecules slow down and their energy levels decrease. This provides more opportunity for the individual molecules to form the maximum number of hydrogen bonds with their neighbours. For this to happen, the water molecules spread out (expand) more. Water is at its most dense at 4°C and, since ice floats, living organisms can survive under the ice. Ice is also a poor **conductor of heat** and helps to insulate the water below, preventing further heat loss.

Biological molecules

KEY SKILLS
C3.1b, IT3.3, WO3

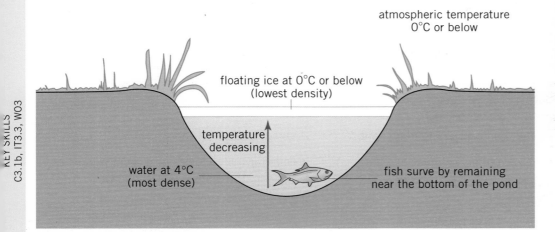

Cross-section through a frozen pond

- Because of its polar nature, water is a good **solvent** for polar molecules, such as sugars and glycerol, and ionic compounds, such as inorganic mineral salts. Here, water molecules are attracted to, and surround, the solute particles, separating them. Once hydrated (dissolved) the solute particles are free to move and react with other chemical substances, also in solution. Water is therefore the medium in which chemical reactions can take place.

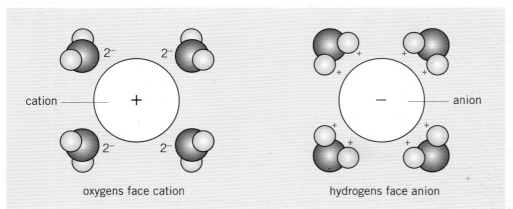

cation

oxygens face cation

anion

hydrogens face anion

Attraction of the polar regions of water to different ions

Chemicals of life: water and minerals

- Non-polar molecules (such as lipids) are **insoluble** (do not dissolve in water). Instead, these molecules tend to get pushed together by the hydrogen bonding between the water molecules that surround them. This property is used, for example, to increase the stability of some proteins and membrane structures.
- Some large molecules have charged areas on their surface that attract water molecules. This layer of water keeps the larger molecules apart and prevents them from joining together and settling out. This is known as a **colloid** suspension (rather like tomato ketchup). Such molecules cause an osmotic effect. The solvent characterisitics make water a good transport medium.

The importance of inorganic salts in living organisms

Inorganic minerals	Function in animals	Function in plants
Calcium (Ca^{2+})	Main constituent of bones, teeth and shells. Muscle contraction and blood clotting. Deficiency leads to rickets and poor clotting.	Translocates carbohydrates and amino acids in phloem. Pectate forms middle lamella. Deficiency leads to stunted growth.
Sodium (Na^+)	Needed for kidneys, nerves and muscles. Sodium helps maintain the osmotic and ion balance across cell membranes and is involved in active transport. Deficiency can cause muscular cramp.	Found inside vacuoles, it is a constituent of cell sap. It maintains cell turgidity. Abundant in all soils, so deficiency is rare.
Potassium (K^+)	Transmission of nerve impulses. Maintains osmotic and ion balance. It is involved in active transport. Potassium is involved in respiration.	Found inside vacuoles, it is a constituent of cell sap. It maintains cell turgidity. Necessary for protein synthesis and is involved in photosynthesis and respiration. Deficiency leads to yellowing of leaves.
Magnesium (Mg^{2+})	Constituent of bones and teeth and is an enzyme activator.	Enzyme activator and is found in chlorophyll. Deficiency leads to yellowing of leaves.
Chlorine (Cl^-)	Helps to transport CO_2 in the blood and helps form hydrochloric acid in the stomach. Maintains osmotic and ion balance. Deficiency can cause muscular cramp.	Chlorine helps to maintain the osmotic and ion balance across cell membranes. Abundant in all soils, so deficiency is rare.
Nitrate (NO_3^-) or Ammonium (NH_4^+)	Nitrogen is a constituent of amino acids, proteins and nucleotides. Some animal hormones require nitrogen for their synthesis.	Required to synthesise amino acids, proteins, nucleotides and chlorophyll. Deficiency leads to yellowing of leaves and stunted growth.
Phosphate (PO_4^{3-}) or Orthophosphate ($H_2PO_4^-$)	Constituents of proteins, nucleotides and ATP. They are involved in respiration phosphorylation and form phospholipids. These ions are important constituents of bones and teeth. Deficiency leads to rickets.	Constituents of proteins, nucleotides and ATP. They are involved in respiration phosphorylation and form phospholipids. Deficiency leads to dull, dark green leaves and stunted growth.

AS Guru™ Biology

Separation of biological molecules

We use various biochemical techniques to separate and identify biological molecules. One simple technique is **paper chromatography**, which is used to separate solutes in a mixture. This process is outlined below:

- draw a pencil line, about 1.5cm from the edge of a piece of absorbent paper
- on the line, mark a number of points, about 2.5cm apart
- place a drop of the solution to be analysed on one of the marked points
- on the other points, put a drop of a substance that you suspect is in the mixture
- let all the spots dry
- place the paper vertically in a tank of solvent (make sure it does not touch the sides), so that the starting line touches the liquid.
- Allow the solvent to travel most of the length of the absorbent paper.
- Mark the position of the **solvent front** and allow the paper to dry.
- If you can't see the **solute fronts** as coloured spots or bands, then you need to find a **locating agent**, which reacts with the solutes to make them visible. Ninhydrin, for example, can be used to develop amino acids.

GURU TIP
Common solvents include water, ethanol (alcohol), ethanoic acid (vinegar) and propanone (acetone). You have to experiment to find a solvent which gives a good separation of the solutes.

Biological molecules

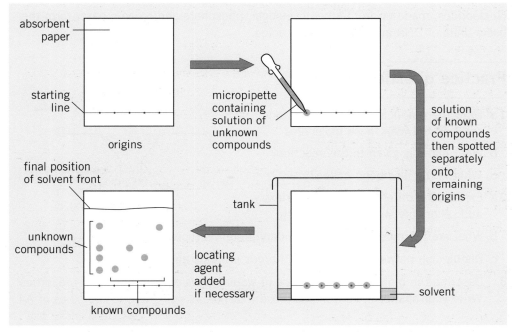

Paper chromatography separates out the component solutes in a chemical

The separation of the solutes depends on their solubility in the solvent. A very soluble solute will travel further along the support medium than a less soluble solute. Some solutes may not move at all if they are insoluble. This technique can tell you how many compounds are in a particular mixture and, by matching against a known pure substance, what the compounds are.

Retardation factor (R_f)
The longer you leave the support medium in the solvent, the further the solute fronts will move. To compare solutes, you need to calculate the **retardation factor (R_f)**, which represents how much they are slowed down. Every compound has its own R_f value in a particular solvent.

$$R_f = \frac{\text{distance travelled by the solute compound}}{\text{distance moved by the solvent front}}$$

Summary

Many biological molecules are formed from units that bond to form polymers, including carbohydrates, proteins, lipids and nucleic acids. The bonds result from condensation reactions (the removal of water molecules).

Because of hydrogen bonding between its molecules and those of other substances, water has many properties which make it important in sustaining life on our planet.

The properties of different forms of carbohydrate depend upon the position of the glycosidic bond, the type of monosaccharide and any cross-branching of units.

Proteins are made up of amino acids held together by peptide bonds. Proteins are either fibrous (insoluble) or globular (soluble). Different bonds (ionic, hydrogen, disulphide and van der Waal's forces) between polypeptides can lead to complex secondary, tertiary and quaternary protein structures.

Lipids are made from fatty acids and glycerol and are hydrophobic. Triglycerides and phospholipids are important forms of lipid.

Nucleotides, made up from a pentose sugar, phosphate and organic base, form the basic units of DNA and other nucleic acids.

Practice questions

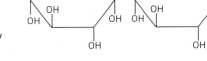

1a Draw the products of a condensation reaction between these molecules:

b What name is given to this reaction?

c What is the new bond called?

2a Explain the difference between primary and secondary structures of protein.

b What are the most common secondary protein structures?

c Distinguish between globular and fibrous proteins and give examples.

3 Describe the biological significance of water's physical and chemical properties. Write your answer in continuous prose (not notes), using one to two sides of A4.

Key skills

Communication, working with others, learning performance and most of the required IT skills can be easily demonstrated by a carefully organised and structured research project. Take the following typical essay-style question as an example:

a Write an illustrated account of the variety of carbohydrates, in living organisms.

b Outline the roles played by carbohydrates in different organisms.

You will be able to find out more about carbohydrates than appears in this book, so carry out some extensive research using the Internet, CD-ROMs and available text resources. Work with others, share information and review your collaborative efforts. Set targets and produce a plan of how these will be met. Then produce a wordprocessed one-page summary, a report for presentation (include overheads/models or produce a PowerPoint presentation) and an extended essay answer. Include appropriate illustrations, website addresses and book references.

Cell Membranes

In this section you will be learning about:

☞ the fluid mosaic model of membrane structure

☞ the components of cell membranes and their functions

☞ how substances pass through cell membranes

☞ the effects of immersing cells in different solutions.

The first section discussed the structure of cell organelles. This section looks at the cell membrane in detail. All cells are surrounded by a surface membrane, which controls the exchange of materials between the inside of the cell and its environment. This includes the cell taking up nutrients from its environment and the removal of excretory products to its surroundings.

The membrane is very thin (about 7nm) and yet is capable of offering strength to the cell as well as precisely controlling the flow of materials. Certain components of cell membranes allow for communication between cells by responding to hormones.

This section shows you how cell membranes are built up from their separate components. It is important that you see how biological molecules have major implications to the structure and function of membranes.

Once you understand the structure of a cell membrane, you will be able to see, more clearly, how materials can enter and leave the cell through it. At KS4, you will have investigated the processes of diffusion, osmosis and active transport and their role in the cellular exchange of materials. Try to remember what you can about these. In this section, these processes are explained in more detail and you should try to understand how they relate to the structural components of the membrane.

Other methods of material exchange are explained in this section. These are facilitated diffusion, phagocytosis, exocytosis and pinocytosis. You may not have come across these before, so you may need to spend a bit of extra time on them.

The structure of cell membranes and the exchange of materials are important concepts. Examiners realise this and often ask examination questions in this area.

Components of cell membranes

The previous section looked at the structure of **phospholipids**. This section looks at this in more detail and the important role that phospholipids play in cell membranes.

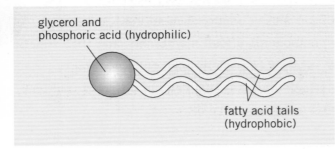

Basic structure of a phospholipid

If you gently introduce a phospholipid onto the surface of water, it spreads out as a layer and the molecules arrange themselves into a sheet. The hydrophilic heads point into the water, since they are attracted by the polar water molecules. The hydrophobic tails stick out of the water, producing a layer of highly organised molecules – rather like synchronised swimming tadpoles.

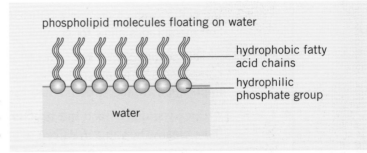

Phospholipids forming a monolayer structure

If you shake a phospholipid with water, it forms a **bilayer** (double layer) as the water surrounds it. All the hydrophilic heads stick outwards and the hydrophobic heads face inwards. This is the underlying structure of cell surface membranes and it has been formed from a bilayer of phospholipid molecules about 7nm thick.

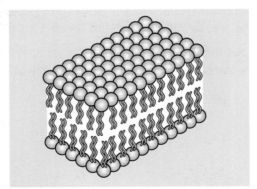

Phospholipid bilayer

Some of the phospholipid tails in the bilayer may be unsaturated (see the previous section). Since these tails have 'kinks' in them, adjacent phospholipids may move further apart. This makes the bilayer more fluid than if all the phospholipid tails were saturated.

There are other structural molecules that are found in cell membranes and are important to the way that the cells function. Use the biological molecules section to revise the properties of these molecules.

- **Cholesterol** is a form of steroid (you don't need to know its structure), with a hydrophilic head and hydrophobic tail. Molecules of cholesterol can fit neatly between molecules of phospholipids. Cholesterol is found in both phospholipid layers.

There are various types of protein found embedded in, or sticking to the surface of the membrane. These can be either fibrous or globular protein types. Some contain channels running through them.

The proteins remain in the membrane because the hydrophobic amino acid R groups sit in the hydrophobic regions of the membrane, and the hydrophilic R groups sit within the hydrophilic parts of the membrane.

- **Glycoproteins** are either fibrous or globular proteins, with a relatively short, glycogen-like chain of carbohydrates attached to the surface that are typically 6–12 units long and are often branched.

- **Glycolipids** are similar to glycoproteins except that the carbohydrate is bonded to a phospholipid molecule.

Intrinsic proteins pass all the way through the membrane.

Extrinsic proteins are found either embedded in the top or lower bilayer, or attached to the upper or lower surface.

Channel proteins are intrinsic and have some sort of passageway running through them.

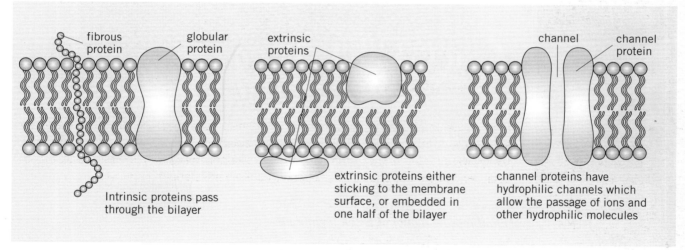

Intrinsic, extrinsic and channel proteins are important in cell membrane structure

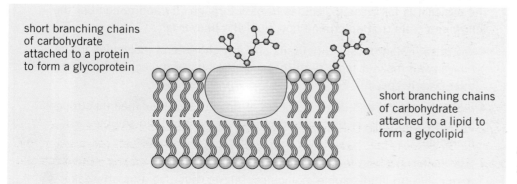

Carbohydrates attach to proteins and lipids in the cell membrane

Cell membranes

The fluid mosaic model

KEY SKILLS
PS3, LP3

The cholesterol, protein, glycoprotein and glycolipid constituents of a cell surface membrane are not static, they are constantly on the move, floating inside the fluid phospholipid bilayer. This is described as a **fluid mosaic** pattern.

The main function of the cell surface membrane is to act as a **selective barrier** between the inside of the cell and its external environment. Most **water-soluble** and polar molecules, such as glucose and ions are prevented from passing through the bilayer because of the hydrophobic tails of the phospholipids. Cholesterol also helps to stop ions, polar molecules and water from leaking through the membrane. These particles are forced to move through special channels formed by carrier proteins and channel proteins. In practice, the cell controls the quantities of these two substances in the bilayer.

The diagram below, shows how the bilayer is organised in the cell surface membrane and how its additional components are arranged.

Although not clearly understood, the functions of glycolipids within the bilayer are thought to be concerned with communication between different cells. Glycolipids are found mainly on the outer surface of nerve cells and within chloroplast membranes.

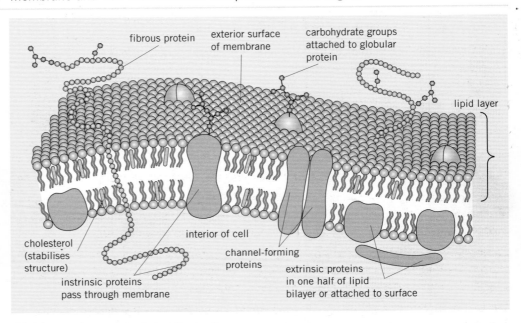

Fluid mosaic model of a cell membrane

GURU TIP
Remember: plant cells have an extra cell wall, made of cellulose. Although this gives the cell extra support, it is permeable to most molecules.

Small **lipid-soluble** and non-polar molecules can generally pass easily between the hydrophobic tails of the bilayer. The more fluid the membrane, the easier it is for these molecules to pass through. Try and remember the following points that determine how fluid the bilayer is:

- any unsaturated phospholipid tails and any cholesterol molecules in the bilayer, will make it more fluid (because the bilayer molecules become more widely spaced)

- more cholesterol and larger numbers of unsaturated tails mean a stronger, but less fluid, bilayer that restricts the movement of lipid-soluble molecules

- less cholesterol and fewer numbers of unsaturated tails means a more permeable (to lipid-soluble molecules) bilayer, but also a much weaker structure, which may collapse, causing the cell membrane to burst.

AS Guru™ Biology

The proteins embedded within the bilayer perform many different functions, as:

- **carrier** and **channel proteins** used to move materials across the cell surface membrane
- **enzymes**, such as helping with the digestion of food in the gut
- **receptors** for different molecules, such as hormones
- a means of **communication** or connection between cells.

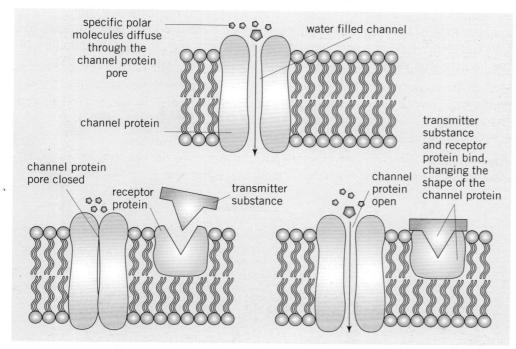

Channel proteins and receptor proteins play an important role in the cell membrane

> Some channel proteins are associated with other **receptor proteins** embedded within the bilayer. For the pore of the channel protein to be open for diffusion, the receptor protein needs to first bind to a specific transmitter substance. This binding process affects the shape of the channel protein, effectively opening the channel.

There are many types of channel protein. Each one is responsible for the transport of a specific polar molecule, or ion, across the bilayer. Each channel protein forms a water-filled tube, through which only a particular particle can diffuse.

Carrier proteins are specific for a particular polar molecule or ion. They continually change, oscillating between one shape and another, perhaps as many as 100 times a second. This helps diffusion of particles across the bilayer.

The carrier protein has a binding site for a particular solute particle, such as glucose. This particle can enter or leave the binding site at random. As the protein changes shape, any attached particle is carried through the bilayer. Some particles will be released on the other side of the bilayer, whilst others may stay attached and be carried back. If there is a higher concentration of solute particles on one side of the bilayer than the other, there will be an increased chance of them binding to the carrier protein on this side and being transported across. Some, of course, will be transported in the opposite direction, but the net movement will be from high to low concentrations.

Carrier proteins transport solutes across the cell membrane

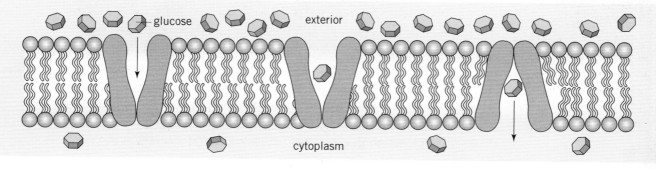

Cell membranes

Cellular transport: diffusion

If someone dropped a stink-bomb in the centre of a room, it wouldn't be long before everyone in the room would be able to smell it, wherever they are. This is an example of **diffusion**. The particles that make up the smell from the stink-bomb spread out in all directions. They move from a region where there is a high concentration of the fragrance particles, to where there is a lower concentration. In effect, they move down a **concentration gradient**, from high to low concentrations of the particle. The steeper the concentration gradient, the faster the rate of diffusion.

GURU TIP

In any question about diffusion look for the high concentration of a particular particle and trace it to the lowest concentration for the same particle. This gives you the direction of movement.

Some particles will move back to where they started. Diffusion is the net movement from high to low concentration.

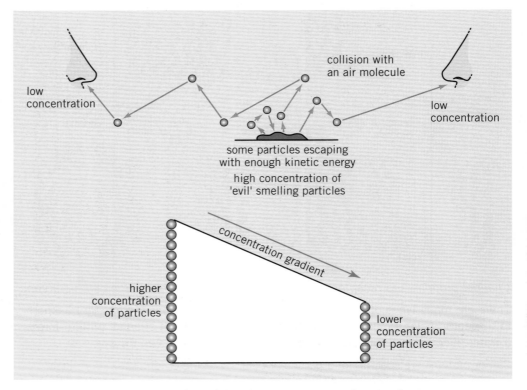

Diffusion is the movement of particles down a concentration gradient

> **Remember:** diffusion is said to have occurred when particles (molecules and ions) in a liquid or gas move from a region of high concentration to a region of low concentration. It is a process of spreading out, which will continue until there is an even concentration of particles throughout.

Kinetic energy is the energy of movement

Diffusion only takes place in liquids and gases because their particles are free to move in random directions with the **kinetic energy** they contain, unlike solids. The higher the kinetic energy of the particles (the warmer they are), the faster the rate of diffusion.

KEY SKILLS
C3.1b, IT3.3, WO3, LP3

Factors that increase the rate of diffusion in the mammalian lung are:

- large surface area
- moist lining – diffusion rate of gases is more efficient if they first dissolve in water
- small diffusion distance – capillary and alveolar cells very thin
- warmth – to increase the kinetic energy of particles
- small size of diffusing particles
- good blood supply and ventilation of alveoli to maintain the diffusion gradient.

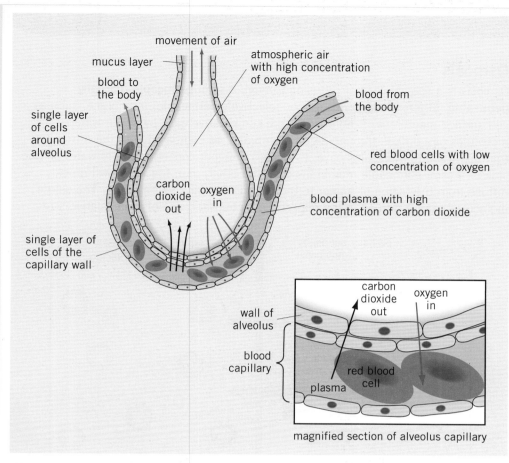

Mammalian lung structure showing features for efficient diffusion

Diffusion accounts for the movement of water, oxygen and carbon dioxide through cell membranes. Oxygen (O_2) is a small, non-polar molecule and easily passes through the spaces between phospholipid molecules, forming the cell membrane bilayer. Water (H_2O) and carbon dioxide (CO_2) on the other hand, consist of polar molecules, which would normally be repelled by the phospholipid bilayer. These molecules however, can pass through the bilayer quickly because of their very small size.

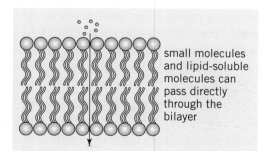

small molecules and lipid-soluble molecules can pass directly through the bilayer

Diffusion of molecules through the bilayer

Large polar molecules (for example, amino acids, glucose, nucleotides), charged ions (such as Na^+ and Cl^-) and some non-polar molecules can't pass through the phospholipid bilayer, either because of their polar nature or their size. The polar molecules are able to pass through the bilayer with the assistance of either channel or carrier proteins. Because diffusion needs to be helped along by these special proteins, the process is known as **facilitated diffusion**.

GURU TIP
The examiners will be looking for ways to test your understanding of basic principles of biology by asking you to relate them to different situations.

Cell membranes

Cellular transport: osmosis

Osmosis

Osmosis is a special case of diffusion through a **semi-permeable** (**partially permeable, differentially permeable** or **selectively permeable**) membrane, such as a cell surface membrane. It relies upon the comparative ease by which water molecules can diffuse through a cell membrane, as compared with the difficulty that many solute molecules have. If there is a higher concentration of water on one side of a semi-permeable membrane than the other, then the water will quickly diffuse from the region of higher to the region of lower concentration. The solute molecules cannot diffuse in the opposite direction because of their relative sizes.

GURU TIP

Find out which of the following terms your examination board uses: (**semi-permeable, partially permeable, differentially permeable** or **selectively permeable membrane**).

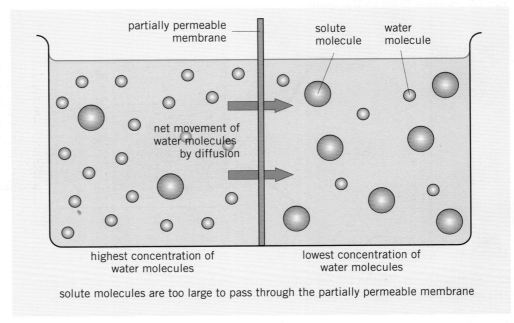

Osmosis is the movement of water molecules through a partially permeable membrane

GURU TV

Osmosis can be a bit tricky to get the hang of, at first. AS Guru™ Biology expert, Mary Jones, sheds some light on this form of transport in an interview on the cells programme.

Water potential

Water potential describes the ability of water molecules to move. Water will always move from a region of high water potential to a region of low water potential by diffusion. Pure water has the highest water potential (highest concentration of water molecules), whereas a solution (water with dissolved solutes) will have a lower water potential. Solute molecules, therefore, reduce water potential. They do this by reducing the ability of water molecules to move. This is a result of attractive forces between the solute and water molecules. The amount of water movement that is lowered is called the **solute potential** and is always a negative value.

Pure water has a water potential of zero and solutions have water potentials of less than this. Water potential is represented by the symbol, Ψ **(psi)** and the units of pressure are kilopascals or megapascals (kPa, MPa). The symbol used for solute potential is Ψ_S.

The pressure exerted on the contents of cells will not only be dependent on the water and solute potentials, but will also be influenced by the pressure exerted by the cell membrane and cell wall on the cell contents. In effect, this is a backwards pressure, trying to squeeze water out of the cell. This pressure is known as **pressure potential** and is shown by the symbol Ψ_P. Pressure potential is normally a positive value.

$$\Psi_{cell} \quad = \quad \Psi_S \quad + \quad \Psi_p$$

water potential of plant cell	=	solute potential	+	pressure potential

Water potential is the ability of water molecules to move

If the cell membrane is not in contact with the cell wall, you can forget about pressure potential, as it will be zero at this point.

Use the following diagrams to explore what happens to plant and animal cells when they are placed in various concentrations of solution. Try and learn them, so that you can re-draw them in the exam.

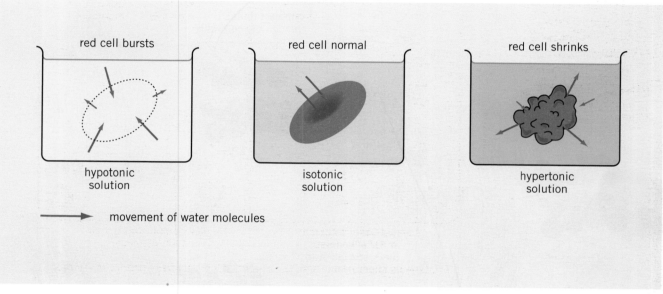

Animal cells in hypotonic, isotonic and hypertonic solutions

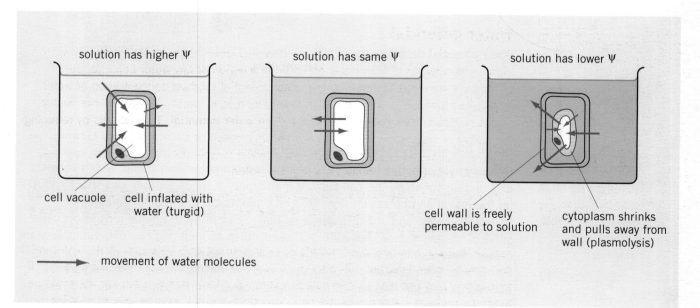

Plant cells in hypotonic, isotonic and hypertonic solutions

Cell membranes

Active transport

The use of energy

Diffusion, facilitated diffusion and osmosis are passive processes where no additional energy (other than the kinetic energy of the particles involved), is needed for them to take place. These processes involve the movement of particles down a concentration gradient. However, in **active transport** energy is needed. This is provided by the **ATP** molecule (you can find out about ATP in the next section), which is produced by cell respiration. Active transport is generally used to move particles against a concentration gradient (from lower to higher concentrations). Like facilitated diffusion, active transport relies on carrier proteins that are specific to a particular particle.

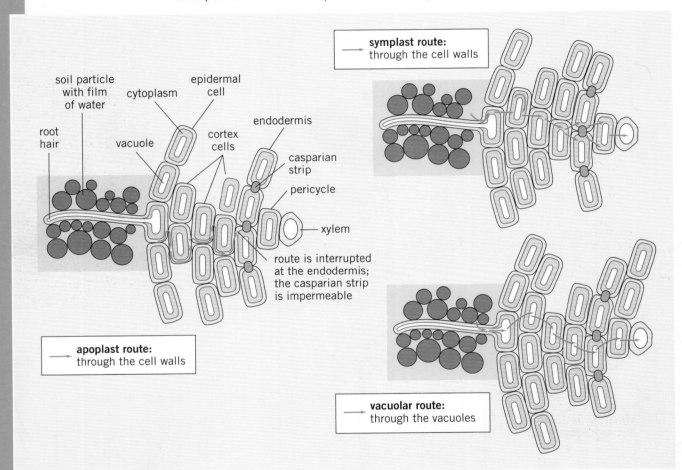

Active transport moves particles against a concentration gradient

One example of active transport involves a **sodium-potassium pump,** which consists of a protein spanning the cell surface membrane. It binds sodium ions (Na^+) and ATP on the inside and potassium ions (K^+) on the outside. The protein splits ATP, releasing energy, and changes shape so that it can move the metal ions to opposite sides of the membrane. For each ATP used, three sodium and two potassium ions are moved. This results in a more positively charged region on one side of the membrane. Both sodium and potassium ions can diffuse through the membrane, however potassium ions can diffuse more quickly than the sodium ions. This results in an even greater charge difference on either side of the membrane. The pump mechanism is most clearly demonstrated by the resting potential of a nerve cell.

outside cell
(low concentration
of molecule, such as
glucose)

membrane ⎯

carrier protein takes
up more molecules
from outside membrane

glucose
molecule

carrier protein
spanning membrane

glucose molecules,
a carrier protein
and ATP attach to the
membrane protein on
the inside of the cell

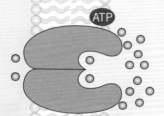

binding of glucose molecules
to protein causes the protein
to change shape (**active
configuration**), so glucose
molecules are now open to
the inside of the membrane
but closed to the outside

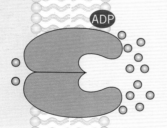

active configuration of the
protein no longer binds
the glucose molecules and
they are released to the
inside of the membrane
by energy released from
the hydrolysis of ATP

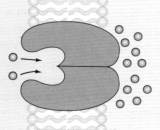

release of the glucose
molecules causes the protein
to revert to its binding
configuration and so it is
available to take up more
glucose from the outside

Try and remember
these three
important points:

- in active transport,
  molecules and ions
  are moved across
  membranes against
  a concentation
  gradient

- energy for active
  transport is
  supplied by ATP

- the carrier proteins,
  spanning the
  membrane, are
  specific for
  particular shapes
  of molecule.

Cell membranes

Carrier proteins actively transport molecules across cell membranes

Phagocytosis and pinocytosis

So far, you have looked at various ways in which individual molecules and ions can pass through the cell surface membrane. Cells also have additional methods of transferring materials, allowing them to take in (**endocytosis**), or excrete (**exocytosis**) large quantities of materials at one time. This is called **bulk transport**.

There are two basic types of endocytosis, depending on how much material is moved: **phagocytosis** for large-scale movement and **pinocytosis** for small-scale movement of materials.

Phagocytosis

Phagocytosis means 'cell eating' and relies upon cells having the ability to stream their cytoplasm in specific directions so that the shape of the cell changes. This also gives the cell the ability to move. Examples of these cells include amoeba (a single-celled animal living in fresh water) and phagocytes (a type of white blood cell).

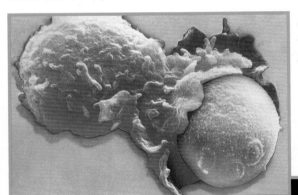

Scanning electron micrograph of a cultured lymphocyte phagocytosing a yeast cell

Scanning electron micrograph of an Amoeba proteus

This section is a good example of where different theories connect and overlap. You will need to understand cell membrane structure, Golgi apparatus, endoplasmic reticulum and lysosomes, as well as excretion and secretion.

KEY SKILLS
C3, IT3, WO3, LP3

Phagocytes are responsible for detecting and destroying foreign organisms, such as bacteria, that have entered the body. The phagocyte forms a flask-shaped depression in its surface membrane (by streaming its cytoplasm around the depression) which surrounds the foreign material. The neck of the flask then gets closed off as a **phagocytic vesicle** that moves inside the cytoplasm of the cell. The foreign particles are then digested by enzymes secreted into the phagocytic vesicle from **lysosomes**, whose membranes fuse with the one surrounding the vesicle. The products of digestion are absorbed into the surrounding cytoplasm by a combination of diffusion, facilitated diffusion, osmosis and active transport. Any indigestible material moves, within its vesicle, back towards the surface membrane. Both membranes now fuse and the contents of the vesicle are excreted. This final stage is the reverse of pinocytic vesicle formation and is also an example of **exocytosis**.

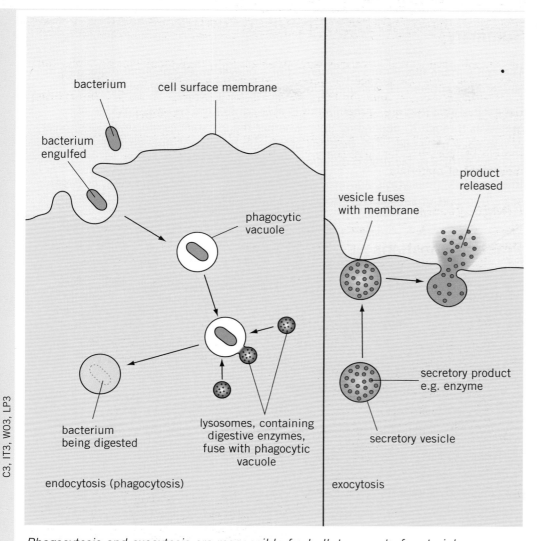

bacterium

cell surface membrane

bacterium engulfed

phagocytic vacuole

product released

vesicle fuses with membrane

bacterium being digested

lysosomes, containing digestive enzymes, fuse with phagocytic vacuole

secretory product e.g. enzyme

secretory vesicle

endocytosis (phagocytosis)

exocytosis

Phagocytosis and exocytosis are responsible for bulk transport of material

GURU TIP
The Golgi apparatus and endoplasmic reticulum are also involved in phagocytosis and pinocytosis.

Cell membranes

Pinocytosis

Pinocytosis means 'cell drinking' and is similar to phagocytosis, only on a much smaller scale. Small **pinocytic channels** are constantly formed at the cell surface membrane. The inner end of each pinocytic channel gets 'pinched off' to form a small vacuole which moves further inside the cell. Smaller vacuoles, in turn, may be made from this vacuole. Pinocytic channels provide a means of getting liquid into the cell. The large quantity of small vacuoles that are made give a large surface area for the absorption of their contents into the cytoplasm.

Exocytosis

Exocytosis is the reverse of endocytosis. It is involved in the bulk removal of materials from the cell. In the example above, waste materials were removed, however this is not always the case. Plant cells, for example, use exocytosis to move cellulose (used to build the cell wall) beyond the surface membrane. In animals, endocrine and exocrine glands produce secretions, such as enzymes (used in digestion) and hormones (used for communication and control within the organism).

GURU WEBSITE
If you want to find out more, or want a bit of revision on the different forms of transport, visit the AS Guru™ Biology website for an interactive guide to them all.

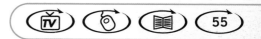

Summary

The cell surface membrane is composed of a bilayer of phospholipid molecules in which other substances are embedded, such as carrier and channel proteins.

Because of its structure, a membrane is able control the two-way passage of materials through it, depending upon whether the substances are small ions, polar or non-polar molecules. The methods of particle movement through a membrane include: diffusion; facilitated diffusion; osmosis and active transport.

Phagocytosis, pinocytosis and exocytosis are methods used for the bulk, or large-scale transport of substances into and out of the cell.

Practice questions

1 The diagram shows the structure of a phospholipid molecule.

   a Label parts A and B.

   b Give three functions of proteins in plasma membranes.

   c Explain why phospholipid molecules form a bilayer in the plasma membrane.

A

B

2a Using particle theory, explain why diffusion occurs.

   b State three features of a membrane that encourage rapid diffusion.

   c Describe what is meant by 'facilitated diffusion'.

3a Osmosis is described as a 'special case' of diffusion. With the help of a simple diagram, explain what this means.

   b Explain why a solution always has a negative water potential.

   c Active transport is used to move particles against a concentration gradient. What supplies the energy for this process?

Key skills

There are plenty of opportunities in this section for further research and presentation of your findings, using a variety of different report formats (word-processed summary or extended reports, PowerPoint presentation or interactive Intranet web page). There are lots of small and large areas that you can find out much more about – choose one. Can you search the Internet, use a CD-ROM, books and other resources to gather appropriate information? Can you look critically at the information sources and extract relevant details? If so, you can easily satisfy IT3.1, IT3.2, IT3.3 (part), and all of C3, LP3 and WO3. Look through the checklist in the introduction to see what you need to do.

If you can show that you can use the water potential formula in this section, you can demonstrate N3.2.

Enzymes

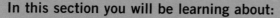

In this section you will be learning about:
- the properties of enzymes and how they are thought to work
- the factors that affect the rate at which enzymes work
- how photosynthesis and respiration are complex chemical processes, controlled by enzymes.

A great deal of chemistry takes place within a cell. Unless all the chemical reactions are controlled and coordinated, there would be complete chaos and the cell would not survive. In KS4, you will have come across the special molecules that act as biological catalysts, speeding up or slowing down reactions, without being chemically involved within the reaction. These important molecules are special globular proteins, called enzymes.

It is important for you to realise that each particular type of globular protein has its own unique and precise three-dimensional shape (depending on the order and number of amino acids in the polypeptide chain). It is this shape that allows the enzyme molecule to control a specific chemical reaction. It is also important to appreciate that a specific enzyme molecule has only one shape and it can, therefore, only control one chemical reaction. This means that every chemical reaction taking place inside the cell must be controlled by a different enzyme, specific for that reaction, if the cell is to exert some control over its activities.

This section reminds you of the many important properties of enzymes and explains how they are thought to work. Try to recall what you know already, before going into detail.

A number of conditions affect the rate at which enzymes operate. Many of these relate to the type of bonding within the polypeptide chain. If you are not sure of what these bonds are, you will need to revise them from the biological molecules section before you can fully understand the explanations given in this one.

If you have understood the various properties of enzymes and their function, take a look at the role of enzymes in the important chemical pathway examples of photosynthesis and respiration, explained later in this section.

What are enzymes?

Enzymes are special, **globular proteins**, which alter the speed of chemical reactions. They are **biological catalysts**. Enzymes usually have a tertiary structure but can also show a quaternary form. In all cases, they have a highly specific, three-dimensional shape that determines how they work.

There is a great deal of chemistry going on, both inside the cell and within the organism as a whole. This is called **metabolism**. Some reactions build up more complex molecules from simpler ones (an **anabolic** reaction), whereas some reactions break down large molecules into simpler ones (a **catabolic** reaction). Each different chemical reaction is controlled by a specific enzyme. There are therefore very large numbers of different enzymes found within living organisms.

Many chemical reactions would not take place without being given some extra energy, perhaps in the form of heat. This is known as the **activation energy**. Enzymes reduce the amount of activation energy needed for a reaction to take place. They help substrate (reactant) molecules to come together and react, to form a new product as in an anabolic reaction, or help a large substrate molecule to break down into simpler products, as in a catabolic reaction. In other words, enzymes help reactions to take place more quickly, at much lower temperatures, than would happen without them.

KEY SKILLS C3.2

GURU TIP
You can find out much more about the structure of proteins in the second sections of this book.

GURU TIP
Enzyme names always end in -ase and the rest of the name generally gives you an idea which substance it works on, for example maltase works on maltose.

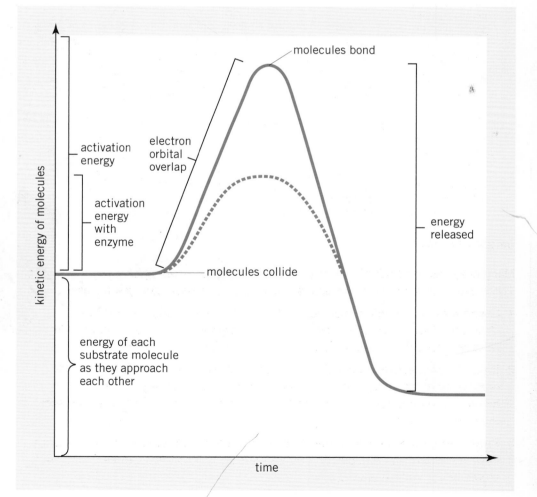

Enzymes lower the activation energy of a reaction

There are two theories that will help you to understand how it is thought that enzymes work: the **lock and key theory** and the **induced fit theory**.

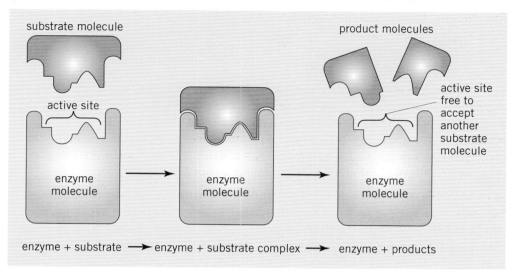

Lock and key theory

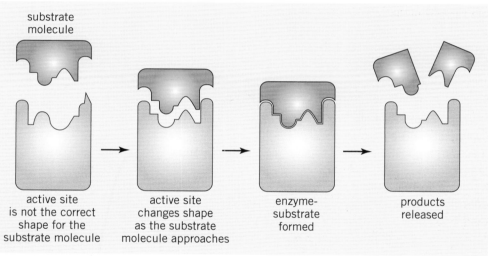

Induced fit theory

Many enzymes need additional components in order to function properly. These extra parts are called **cofactors**. There are three basic groups of these:

1 **Coenzymes** are organic molecules that often contain vitamins within their structure. Coenzymes bind loosely with the enzyme molecule and move away when the reaction is completed.

2 Some enzyme activators consist of inorganic **metal ions**, such as Fe^{2+} in the enzyme catalase. These alter the charge inside the active site of the enzyme, helping it to bond temporarily with the substrate molecules, and this increases the rate of reaction.

3 **Prosthetic** groups are coenzymes that bind to the enzyme molecule permanently. FAD (flavin adenine dinucleotide) is an example of this. FAD carries hydrogen atoms using oxidase enzymes.

Coenzyme NAD (nicotinamide adenine dinucleotide) transfers hydrogen atoms in a series of reactions using a number of dehydrogenase enzymes. NAD is involved in the process of oxidative phosphorylation, which is part of the respiration metabolic pathway (later in this section).

Enzymes

Factors that affect enzyme activity

GURU TIP

Make sure that you really get to grips with these factors. They are a favourite with examiners.

GURU TIP

The rate of enzyme action is measured by the amount of substrate changed, or the amount of product formed, during a period of time.

KEY SKILLS
C3.2, IT3.2, N3

When you cook an egg, you are denaturing globular proteins and forming fibrous proteins.

The following section explores five conditions that affect the rate at which enzymes work. Make sure that you keep the following facts about enzymes in your mind:

- enzymes are globular proteins
- they act as biological catalysts
- they are specific for a particular chemical reaction
- they are not destroyed by the reactions they catalyse
- they reduce the activation energy in a chemical reaction.

The effect of temperature

If you increase the temperature of an enzyme in solution, its kinetic energy (energy of movement) will also increase. This causes the enzyme to move more quickly, increasing its chances of colliding with its specific substrate molecules. Therefore, increasing the temperature increases the rate of the enzyme-controlled reaction, up to a point. At a particular temperature (the **optimum temperature**), the enzyme will work at its fastest rate. Beyond this, any further increase in temperature will cause the enzyme molecule itself to vibrate so much that the bonds (hydrogen, disulphide and ionic bonds, and van der Waal's forces – biological molecules section) holding the three-dimensional structure together distort or even break. This changes the shape of the enzyme and its active site. This makes the enzyme less effective at forming the enzyme-substrate complex and the rate of reaction falls. If an enzyme is heated too strongly, the bonds holding the molecule together break permanently. The enzyme loses its shape and reverts to a fibrous structure. At this point, the enzyme is **denatured** and will no longer function.

Homeothermic animals (mammals and birds) maintain a constant body temperature (37 °C in humans) so that their enzymes work at the optimum rate. Poikilothermic animals rely upon the temperature of their surroundings. This is why an alligator, for example, will be rather sluggish after a cold night (losing the heat it has built up during the day to its surroundings), but after bathing in the sun for some time it will become much more active. This is because the increase in body temperature is matched with an increase in enzyme activity, particularly those involved in energy production through the process of respiration.

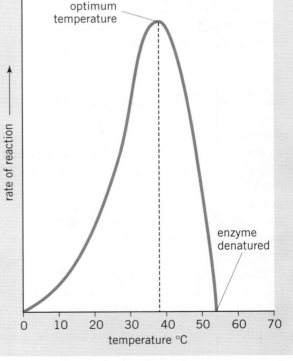

The effect of temperature on enzyme activity

The effect of pH

As with temperature, each enzyme has its own optimum pH. If the pH changes much from this optimum, the charges on the active site of the enzyme or the substrate are affected. This slows down (or stops altogether) the rate of the enzyme-substrate complex formation. At extreme levels of pH, the enzyme's molecular bonds (including those forming the structure of the active site) are broken and the enzyme becomes irreversibly denatured.

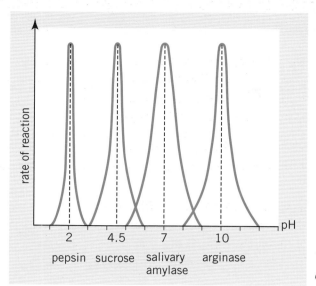

Most human enzymes are found inside cells and have an optimum pH of 7.3-7.4 (slightly alkaline). Peptidase, found working in the acidic environment of the stomach has an optimum pH of 2.4 (highly acidic).

The effect of pH on enzyme activity

The effect of enzyme concentration

At low enzyme concentration, there are more substrate molecules available than there are enzyme active sites. Increasing the number of active sites by increasing the number of enzyme molecules will therefore increase the rate of reaction. The rate of reaction is found to be proportional to the enzyme concentration. As the enzyme concentration is doubled, the reaction rate is doubled.

Eventually, increasing the enzyme concentration will have no effect on the rate of reaction. At this point, it is the numbers of available substrate molecules that are the **limiting factor**.

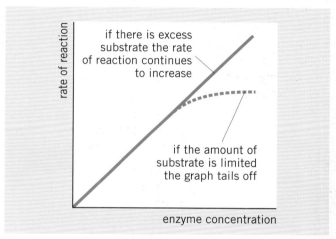

The effect of enzyme concentration on enzyme activity

The effect of substrate concentration

At low substrate concentration, the enzyme-controlled reaction proceeds slowly. This is because there are more active sites available than substrate molecules. As substrate concentration increases, the rate of reaction increases as more substrate molecules are turned into products. At a certain point, there is no further increase in the rate of reaction with the increase in substrate concentration. This is because all the active sites are occupied by substrate molecules. The enzyme concentration has now become the limiting factor. The rate of reaction now depends upon the 'turnover rate' of the enzyme, that is, how many substrate molecules one molecule of enzyme can work on, in one second. The turnover rate of the enzyme catalase, for example, is over five million substrate molecules (hydrogen peroxide) broken down (to water and oxygen) in one second.

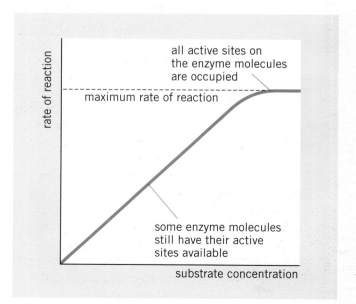

The effect of substrate concentration on enzyme activity

Enzymes

Enzyme inhibition

KEY SKILLS
C3.2, IT3.2, N3

The effect of inhibitors

Enzyme inhibitors are used to slow down chemical reactions, when necessary, inside a cell. For example, when sufficient products have been formed. Inhibition may be reversible or irreversible.

Competitive reversible inhibitors consist of molecules that are structurally very similar to an enzyme's normal substrate molecule. These molecules compete for the active site in the enzyme, but leave after a time without forming a product. While the inhibitor is in the active site, the enzyme cannot form products from its normal substrate and this slows down the reaction. The higher the inhibitor concentration, the slower the rate of reaction.

If you swallowed methanol (a form of alcohol, which in itself is not toxic), you would be in danger of going blind. Methanol is converted to formaldehyde, which is highly toxic, by an enzyme found inside your cells. A treatment for methanol poisoning would be to give you ethanol (the drinking variety of alcohol). This is structurally similar to methanol and competes for the enzyme active sites. This slows the reaction down and results in the slower production of formaldehyde, which in turn, will cause less damage to you. Don't try this at home!

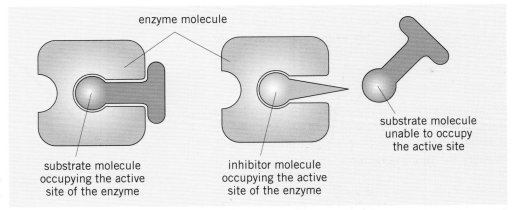

Competitive reversible inhibition

enzyme molecule

substrate molecule occupying the active site of the enzyme

inhibitor molecule occupying the active site of the enzyme

substrate molecule unable to occupy the active site

Non-competitive reversible inhibitors react with the enzymes, but not at their active sites (this is known as an **allosteric enzyme**). The inhibitor molecule binds into an **allosteric site** and this causes the enzyme molecule and the active site to distort in shape. This stops substrate molecules from entering the active site and in turn, stops the formation of products until the inhibitor leaves.

> Cells need enzymes to control chemical reactions but they also need to control the enzymes themselves.

Non-competitive reversible inhibition

enzyme molecule

substrate molecule still fits the active site but not in a way that allows the reaction to proceed

substrate molecule occupying the active site of the enzyme

inhibitor molecule attached to enzyme molecule

enzyme molecule shape is changed due to presence of inhibitor molecule

End-product inhibition

The activity of allosteric enzymes can be affected by compounds binding to the allosteric sites and either speed up (**allosteric activator**) or slow down (**allosteric inhibitor**) the rate of reaction. This can be used to control metabolic pathways consisting of linked chemical reactions, where the end product can be used to switch off the process. This is known as **end-product inhibition**.

GURU TIP
As the concentration of inhibitor increases, the rate of reaction decreases.

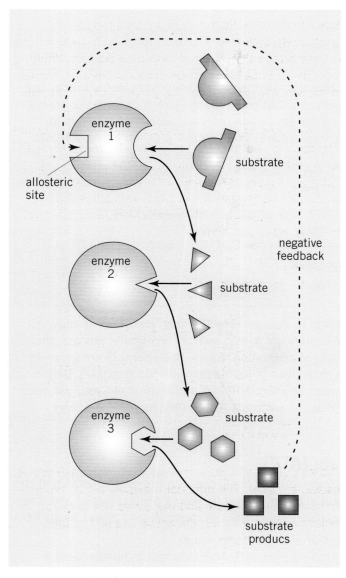

End-product inhibition

GURU TIP
There are many metabolic pathways that show similar enzyme-controlled chemical reactions, like this. Look up photosynthesis and respiration, later in this section, for more details.

Irreversible inhibitors

Very small concentrations of metal ions, such as mercury, silver and arsenic can combine permanently with the active sites of some enzyme molecules. This stops them from functioning altogether.

Photosynthesis and enzyme control

As we know, all living organisms require energy in order to survive. This energy allows an organism to carry out all of the seven life processes that you will remember from your KS4 studies: **movement**, **reproduction**, **sensitivity**, **growth**, **respiration**, **excretion** and **nutrition**. You will find reference to these, and the need for energy to make them possible, throughout this book.

> **Remember:** MRS NERG or MRS GREN?

Organisms either get their energy from light or from chemical potential energy (eats food!). **Autotrophs** are organisms, such as green plants, which can use the energy in light, an inorganic source of carbon (in carbon dioxide), water and minerals to manufacture all the materials they require. This process is called **photosynthesis**. You came across the general equation for photosynthesis in KS4. That was just a summary – photosynthesis actually consists of a complex series of chemical stages, each controlled by separate **enzymes**.

$$6CO_2 + 6H_2O \xrightarrow[\text{in the presence of chlorophyll}]{\text{light energy}} C_6H_{12}O_6 + 6O_2$$

GURU TIP
Since hexose sugars and starch are often formed in the process of photosynthesis, glucose ($C_6H_{12}O_6$) is often shown in the equation.

Autotrophs cannot use the energy in sunlight directly, they 'lock it up' in organic molecules such as glucose. When energy is required for a particular process, this energy is released in the process of **respiration**. Again, this energy is not generally used directly, but is stored in molecules of ATP (adenosine triphosphate). Before you can understand the processes of photosynthesis and respiration, you first need to understand more about ATP as the molecule responsible for moving energy around a cell.

Adenosine triphosphate (ATP)

ATP is a nucleotide (you can find out more about these in the second section of this book) that consists of adenine (an organic base), ribose (a pentose sugar) and three phosphate groups. ATP is a relatively small, soluble molecule and moves easily from one part of a cell to another.

When a phosphate group is removed from ATP, ADP (adenosine diphosphate) is formed and a great deal of energy is released (30.6kJ/mol). The energy comes partly from breaking the bond and partly from changes to the chemical potential energy within the molecule as a whole. If a further phosphate group is removed from ADP, **AMP** (**adenosine monophosphate**) is formed and a further 30.6kJ/mol of energy is released. If AMP loses its final phosphate group to form adenosine, a smaller amount of energy is released (13.8kJ/mol). In reality, rarely does the process get to the AMP and adenosine stages within cells.

These reactions are all reversible, so ADP can be built up into ATP with the addition of a phosphate group and sufficient energy (coming from respiration, for example). Therefore ADP can take up energy and ATP can release energy. This way, ATP operates as the 'energy currency' for a cell, transferring it to wherever it is required.

It is important that when energy is released from some process in the cell, it is carried to where it is actually needed. This is done by ATP.

AS Guru™ Biology

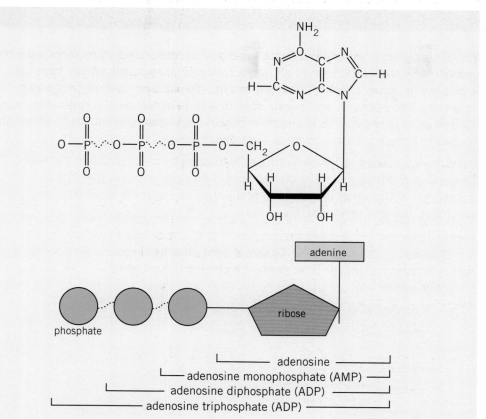

Structure of ATP

ATP production generally takes place in the thylakoid membranes of chloroplasts or inside the inner membrane of mitochondria (see cell structure, function and organisation section). Energy is stored as an electrical potential difference of hydrogen ions (H^+) on either side of the membrane, which is impermeable to these ions. Channel proteins embedded within the membrane allow the hydrogen ions to diffuse through them along their concentration gradient (see the previous section). Part of the channel protein acts as an enzyme, which can make ATP from ADP and a spare phosphate group (Pi) and is called **ATP synthetase**. For every three hydrogen ions passing through the channel protein, one molecule of ATP is produced.

ATP production

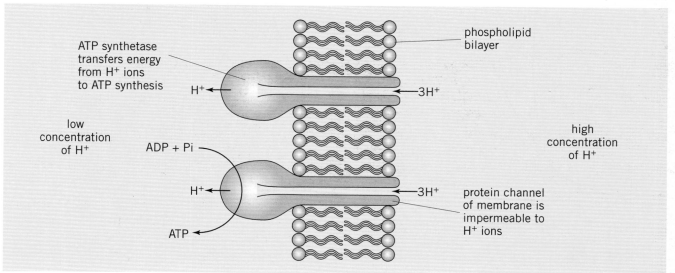

Enzymes

Stages of photosynthesis

You can think of photosynthesis as a two-stage process, one that requires light and another, which doesn't. In the light stage, chlorophyll traps the energy from light and converts it into chemical potential energy. This is then used to split water molecules into hydrogen and oxygen. The oxygen is either used in respiration, or evolved as oxygen gas. The hydrogen enters the light-independent (dark) phase and is used to reduce carbon dioxide to form carbohydrate.

Light energy needed for photosynthesis, is trapped by photosynthetic pigments. These absorb different wavelengths of light. The most common pigment is chlorophyll *a*. Use other resources to find out more about the absorption and action spectra on the photosynthetic pigments.

Pigment	Colour of light absorbed
chlorophyll:	
chlorophyll *a*	yellow - green
chlorophyll *b*	blue - green
carotenoids:	
carotenes	orange
xanthophylls	yellow

Photosynthetic pigments and the coloured light absorbed

When light energy falls on chlorophyll *a*, the electrons within the molecules become excited and are raised to a higher energy level. These electrons have sufficient energy to leave the molecule and are picked up by **electron acceptor** molecules.

The photosynthetic pigments are arranged inside the **thylakoid membranes** of **chloroplasts**. A cluster of pigment molecules is called a **photosystem**. Each photosystem is made up of a large number of **accessory** pigment molecules (some forms of chlorophyll *a*, chlorophyll *b* and carotenoids) surrounding a single **primary** pigment molecule (chlorophyll *a*), called the **reaction centre**. The primary pigment molecule can be one of two types depending on the peak wavelength of light absorbed – either 680nm or 700nm.

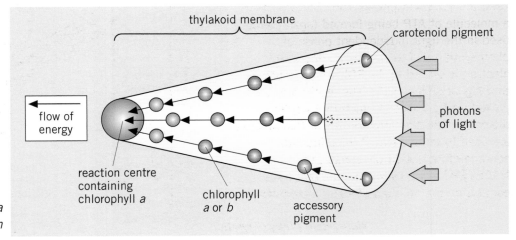

Pigments in a photosystem

In the **light-dependent stage of photosynthesis**, light is used as a source of energy, and water as a source of electrons and hydrogen ions (H^+). These electrons and hydrogen ions are used in later reactions. There are three parts to the light-dependent stage.

Cyclic photophosphorylation only involves **photosystem I (PSI)**. As an excited electron leaves the chlorophyll *a* molecule, it is picked up by an **electron acceptor** molecule. This passes the electron through a further chain of **electron carrier** molecules, eventually returning it, in a reduced energy state, to the chlorophyll *a*. As the electron is passed down the **photophosphorylation** chain of electron carriers, enough energy is released from the electron to form one molecule of ATP. The ATP is used in the light independent reactions of photosynthesis.

Chlorophyll *a* is used in both photosystems but PSI uses one with an absorption peak at 700nm and PSII uses one with a peak at 680nm.

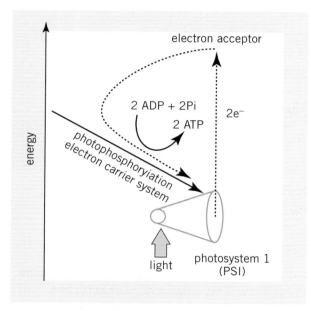

Cyclic phosphorylation

The original electron is not lost from the chlorophyll *a* (it has just been 'borrowed') and is free to become excited again with the addition of more light energy collected by the photosystem.

The electrons lost from PSII are replaced by those from **photolysis**.

Non-cyclic phosphorylation involves both photosystems (PSI and II). The light energy taken up by PSII is used to excite an electron, which leaves the chlorophyll *a* molecule and is picked up by an electron acceptor molecule. This then passes the electron down the same photophosphorylation chain of electron carriers used in cyclic phosphorylation. This results in a molecule of ATP being formed (again being used in the light-independent phase of photosynthesis).The slightly de-energised electron is picked up by PSI, which boosts it to a higher energy state. The electron is then picked up by a second electron acceptor molecule. The electrons formed in this process are used to reduce NADP (nicotinamide adenine dinucleotide phosphate) to $NADP + 2H^+$ using hydrogen ions that have come from the splitting of water molecules.

Non-cyclic phosphorylation

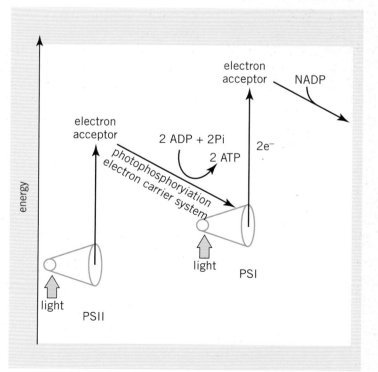

Enzymes

Stages of photosynthesis

Photosystem II contains a water-splitting enzyme. This enzyme is involved in the process of **photolysis**. Photolysis results in the splitting of water molecules into hydrogen ions, free electrons and oxygen:

The hydrogen ions, formed from photolysis, combine with electrons from photosystem I to reduce NADP to $NADP + 2H^+$. This is used in the light independent phase of photosynthesis, along with the ATP formed by phosphorylation.

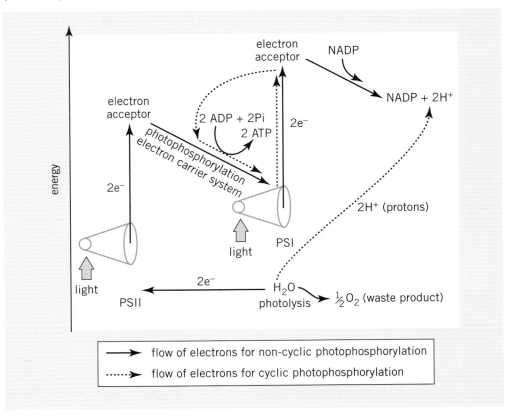

Summary of the light-dependent stage of photosynthesis.

Remember: photophosphorylation, photolysis and the reduction of NADP all take place inside the thylakoids of the chloroplast.

The light-independent phase of photosynthesis takes place in the **stroma** of the chloroplast. This stage will continue whether light is present or not.

Most photosynthetic organisms use the enzyme **ribulose bisphosphate carboxylase** (**RuBisco**) to join carbon dioxide to a 5-carbon molecule of **ribulose bisphoshate** (**RuBP**). This forms an unstable molecule that splits into two, more stable, 3-carbon molecules of **glycerate 3-phosphate** (**GP**). With the addition of ATP and $NADP + 2H^+$, formed from the light dependent phase, GP is converted into a 3-carbon molecule of **triose phosphate** which is a triose sugar and simple carbohydrate. Some of the triose phosphate is converted into hexose sugars, starch and cellulose, whilst some is converted into lipids and amino acids. The majority of triose phosphate, however, is converted back to ribulose bisphosphate, so that the cycle (Calvin cycle) can begin again. This final stage requires a further molecule of ATP.

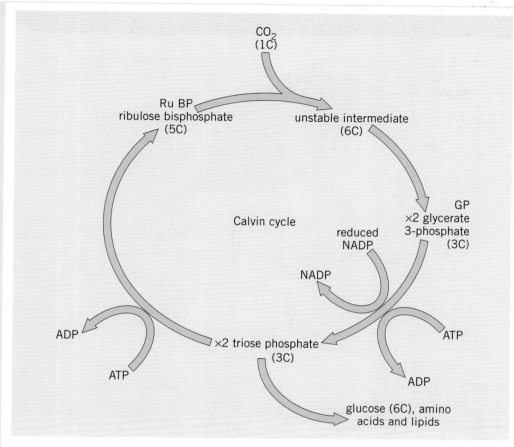

Light-independent stage of photosynthesis

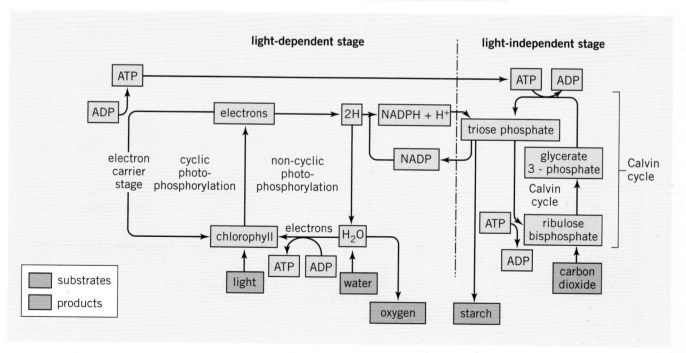

Summary of photosynthesis

Enzymes

Leaf structure and function

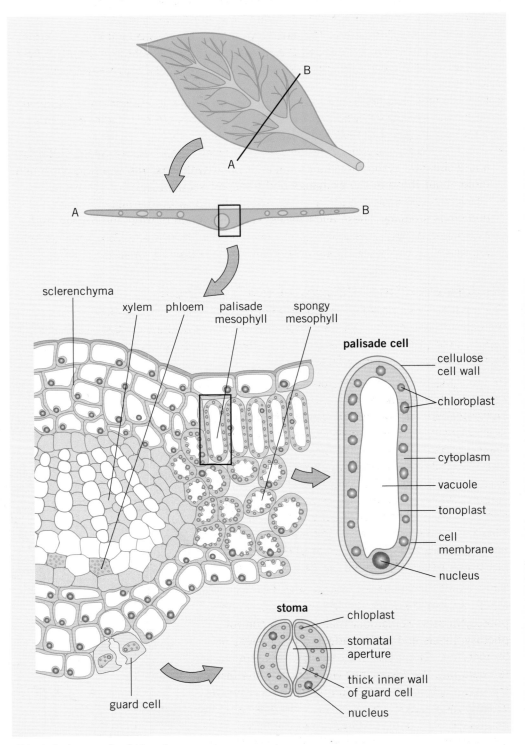

Dicotyledonous leaf structure

The main site of photosynthesis is in the palisade layer. Here, the column-shaped cells are tightly packed together and contain many chloroplasts. Some photosynthesis takes place in the spongy mesophyll layer and guard cells since they also contain chloroplasts.

Leaves have become adapted to increase the efficiency of absorption of light energy.

- The leaf offers a large surface area for light absorption and is capable of movement in order to track the passage of the sun during the day.
- The leaf is very thin so that the lower layers can also absorb light energy. Gases can also diffuse quickly through the leaf, particularly the spongy mesophyll layer.
- The waxy cuticle and epidermis are transparent, allowing light to enter the lower layers.
- Most chloroplasts are found inside the palisade cells and are generally arranged so that the photosynthetic pigments in the thylakoids are perpendicular to the direction of light, for maximum energy absorption.
- The leaf has a good transportation system for water (required for photosynthesis) and a separate system for the products of photosynthesis. You will find out much more about these in the exchange and transport section.
- The lower epidermis (and to a lesser extent the upper epidermis) contains numerous stomata. These allow gases, such as carbon dioxide and oxygen, to be exchanged between the inside of the leaf and the atmosphere. Stomata also help the movement of water through the plant in the transpiration stream.

Factors affecting the rate of photosynthesis

The main factors affecting the rate of photosynthesis are: light intensity; temperature and carbon dioxide concentration. If any of these factors are in short supply, then they will limit the amount of photosynthesis taking place. They are therefore known as **limiting factors**.

In low light conditions, plants photosynthesise slowly. As the **light intensity** increases, the rate of photosynthesis increases in proportion to it. As the light intensity continues to be increased, a point is reached where the rate of increase photosynthesis begins to fall and no longer increases proportionally. Here, one or more of the other factors (for example, carbon dioxide concentration) is beginning to limit the rate of photosynthesis. Eventually, a stage is reached where no further increase in light intensity will cause any further increase in the rate of photosynthesis – it has been completely limited by the other factors.

Similar patterns are shown for the affect of temperature and carbon dioxide concentration on photosynthesis. Remember that enzymes control photosynthesis and will be affected by changes in temperature. The closer they are to their optimum temperature, the faster they will work and the faster the rate of photosynthesis. Similarly, the greater the availability of carbon dioxide, the faster the carbon will be incorporated into carbohydrate in the light-independent phase of photosynthesis. These are both limiting factors as their values increase but they will show the same graphical trend as the affect of light intensity. Eventually the other limiting factors will have an effect.

IT3.2, N3, PS3

GURU TIP
Examiners love interpretation questions, where you are given a graph and asked to explain it.

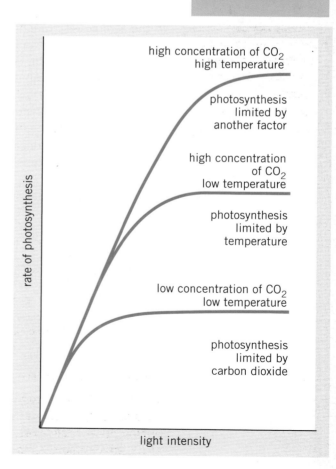

Enzymes

Respiration and enzyme control

KEY SKILLS
IT3, N3, WO3, LP3

GURU TIP
Don't forget that green plants also carry out respiration.

Photosynthesis results in potential chemical energy being locked up in fuels, such as carbohydrates and lipids. Living organisms however gradually release the energy in a series of enzyme-controlled reactions that we call respiration. This series of reactions gradually transfers energy from the fuel into ATP and heat.

There are two forms of respiration: **aerobic** (requiring oxygen) and **anaerobic** (not requiring oxygen) respiration. Aerobic respiration releases much more energy and therefore produces more molecules of ATP than anaerobic respiration.

You should remember from KS4 biology:

The general symbol equation for aerobic respiration:

$$C_6H_{12}O_6 + 6O_2 \rightarrow 6CO_2 + 6H_2O + energy$$

This is the exact opposite of photosynthesis. The energy released is **2880 kJ**.

The general word equation for anaerobic respiration in yeast:

glucose ➜ ethanol + carbon dioxide + energy

The glucose is partially broken down to ethanol, releasing **210 kJ**.

The general word equation for anaerobic respiration in humans:

glucose ➜ lactic acid + energy

Glucose is broken down to **lactic acid**. The energy released is **150 kJ**.

In the aerobic respiration reaction, four high-energy, electrons are removed from each carbon atom in the fuel (glucose in this example) to provide the energy for ATP production. This, again, is exactly the opposite of photosynthesis. The processes of photosynthesis and respiration are compared below. In photosynthesis, the electron donor molecule to the carbon is $NADP+2H^+$ (reduced NADP). In aerobic respiration, NAD (nicotinamide adenine dinucleotide), accepts high energy electrons from carbon, forming reduced NAD ($NAD+2H^+$). Reduced NAD contains a great deal of energy which can be released (and ATP molecules formed) by passing electrons on (through a series of steps) to oxygen.

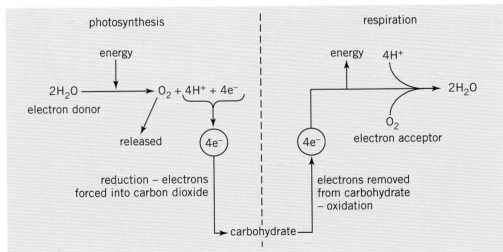

Comparing photosynthesis and respiration

There are four stages involved in aerobic respiration: **glycolysis**, the **link reaction**, the **Krebs cycle** and **oxidative phosphorylation**.

Glycolysis

Glycolysis consists of ten enzyme-controlled reactions, which activate hexose (6-carbon) molecules, such as glucose, and split them into two 3-carbon molecules.

The first stage is known as phosphorylation, using the energy and phosphate groups from two ATP molecules. Molecules of glucose form **hexose phosphate** (one phosphate group) and then **hexose bisphosphate** (two phosphate groups).

In the second stage of glycolysis, the hexose bisphosphate (6 carbons) breaks down into two, 3-carbon molecules of triose phosphate. The triose phosphate is then broken down, through various intermediate stages, into pyruvate (3-carbon) molecules. Between the triose phosphate and pyruvate stages, each molecule of triose phosphate results in the release of sufficient energy to form two ATP molecules and two hydrogen atoms, which are picked up by NAD and reduced to $NAD+2H^+$.

Stages of glycolysis

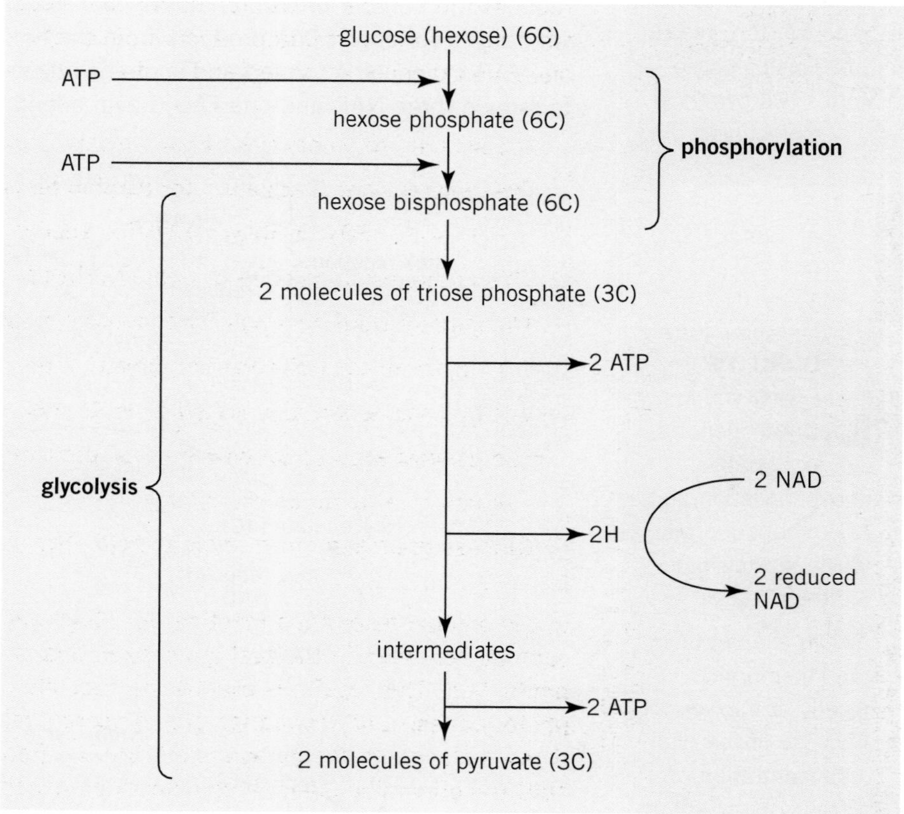

glucose (hexose) (6C)

ATP ⟶

hexose phosphate (6C)

ATP ⟶

⎫ **phosphorylation**

hexose bisphosphate (6C)

2 molecules of triose phosphate (3C)

⟶ 2 ATP

2 NAD

glycolysis ⟶ 2H

2 reduced NAD

intermediates

⟶ 2 ATP

2 molecules of pyruvate (3C)

The link reaction

When **pyruvate** molecules enter mitochondria, they are **decarboxylated** (a molecule of carbon dioxide is removed). The resulting compound is combined with **coenzyme A** to give **acetyl coenzyme A** (**acetyl CoA**) – a 2-carbon molecule. This is the link reaction, resulting in the removal of one molecule of carbon dioxide, the formation of 2-carbon acetyl CoA and the release of 2 hydrogen atoms (picked up by NAD to form reduced NAD – $NAD+2H^+$). This takes place in the matrix of mitochondria.

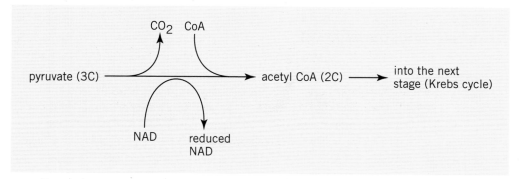

CO_2 CoA

pyruvate (3C) ⟶ acetyl CoA (2C) ⟶ into the next stage (Krebs cycle)

NAD reduced NAD

The link reaction

Enzymes

Respiration and enzyme control

GURU TIP

For one glucose molecule, the Krebs cycle will turn twice, because glucose forms **two** pyruvate molecules.

GURU TIP

Revise what you know about mitochondria, as oxidative phosphorylation and ATP production take place in their inner membranes.

The space between the membranes is usually at a lower pH because of the high concentration of hydrogen ions formed by the electron transport chain.

The Krebs cycle

Krebs cycle takes place in the matrix of mitochondria, combining a 2-carbon acetyl coenzyme A with oxaloacetate (4-carbon) to produce a 6-carbon citrate. The original coenzyme A is released back to the link reaction. The rest of the Krebs cycle reduces the number of carbon atoms (two at a time), back to oxaloacetate, which is used again for acetyl CoA products from the link reaction. In one turn of the cycle, one ATP molecule is formed and some hydrogen atoms are released, which are used to reduce three NAD and one FAD (flavin adenine dinucleotide) molecules.

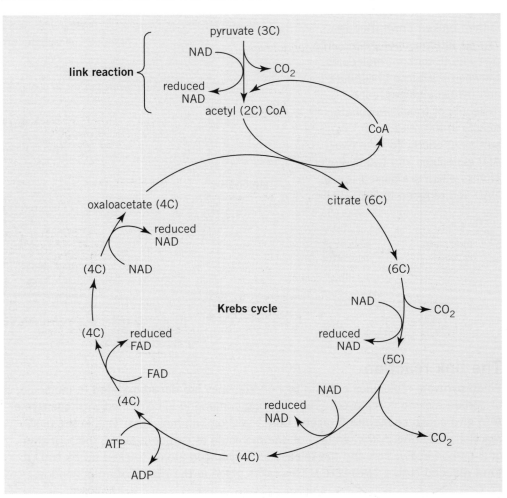

The link reaction and Krebs cycle

Oxidative phosphorylation

Most energy from respiration is released at this final stage of respiration. NAD and FAD carry hydrogen ions and electrons at high energy levels. In oxidative phosphorylation, the hydrogen ions combine with oxygen to form water, and the electron's energy is gradually released through an **electron transport chain** forming ATP molecules. One molecule of reduced NAD forms three ATP molecules. One molecule of reduced FAD forms two ATP molecules (FAD carries electrons at a lower energy level than NAD, so it enters the electron transport chain later). There are particular points at which sufficient energy is released from the electron transport chain to form a molecule of ATP.

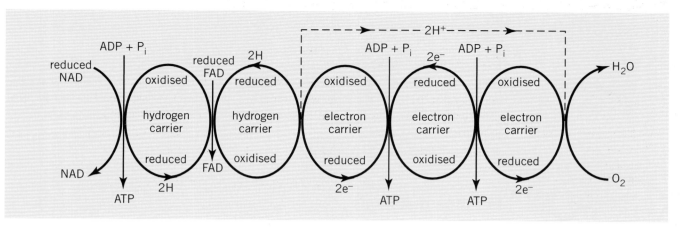

Oxidative phosphorylation and energy levels

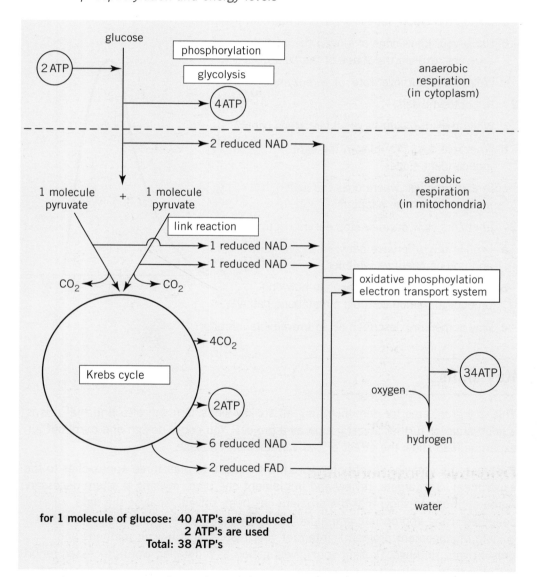

GURU TIP
Without oxygen the majority of ATP production will not take place. The removal of hydrogen ions (by combining them with oxygen to form water), allows more hydrogen ions to be released from the electron carrier system. This is what 'drives' the process.

GURU TIP
Two pyruvate molecules are formed from one glucose molecule in glycolysis and each of these will go through the link reaction, Krebs cycle and oxidative phosphorylation separately.

for 1 molecule of glucose: 40 ATP's are produced
2 ATP's are used
Total: 38 ATP's

Summary of respiration

Enzymes

Summary

You have learnt that enzymes are globular proteins that control chemical reactions. They catalyse reactions by lowering the activation energy. Each enzyme matches its active site to a specific substrate, so anything that alters the shape of an active site (pH, temperature, non-competitive inhibitors) will slow down the reaction rate.

Cell activity is controlled by specific enzymes for each reaction and the enzymes are controlled with inhibitors.

Practice questions

1 The graph shows the effect of temperature on an enzyme-controlled reaction.

   a Sketch the graph, labelling the optimum temperature.

   b Using your knowledge of kinetic theory and enzymes, explain the shape of the curve.

   c What is the 'turnover rate' of an enzyme?

2 In photosynthesis:

   a When does the light-independent stage occur?

   b What role does CO_2 play in the light-independent stage?

   c In a chloroplast, where does the light-independent stage happen?

3 Inhibitors slow down or stop enzyme activity.

   a Explain the difference between a competitive and non-competitive inhibitor.

   b Using a simple diagram, explain how end-product inhibition controls a metabolic pathway.

   c Why is mercury described as an irreversible inhibitor?

(Graph: y-axis labelled "rate of reaction", x-axis labelled "temperature °C" with markings 0, 10, 20, 30, 40, 50, 60, 70. A curve rises to a peak around 40°C then falls sharply to the x-axis near 55°C, with an arrow labelled "enzyme denatured".)

Key skills

This is a great area for demonstrating all six key skills. You will also find that this is a good area for further investigation as a project. You could design and carry out an experiment to show the effect of two variables on catalase.

Research the catalase reaction and work with others to plan three approaches to the experiment. Set targets, review and implement one plan, revising it where necessary. Agree methods to show that the problem has been solved. Analyse the data collected using appropriate graphical and statistical methods (using a spreadsheet or data management program). Interpret your results, justify your methods and present your findings. Present your investigation as a detailed word-processed report with the data displayed in a spreadsheet format and charts, graphs and numerical analysis. Then either produce a PowerPoint demonstration or a one page abstract detailing what you started out to do, your approach to the problem and your main findings. This will fulfill the requirements for all of C3, N3, IT3, WO3, LP3 and PS3 key skills.

DNA

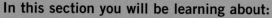

In this section you will be learning about:

☞ the nature of the genetic code and how it is used to manufacture specific proteins inside the cell

☞ how knowledge and understanding of DNA has led to the development of genetic engineering techniques

☞ how these techniques are used in medicine and other fields.

This section explains how all cell activities are controlled by the information stored inside the nucleus. The section on biological molecules looks more closely at the structure of DNA.

The nucleus of a cell contains DNA (deoxyribose nucleic acid), which gives cells a control mechanism. The previous section shows you how enzymes are important in controlling chemical reactions in living organisms. Enzymes are made from proteins, and the DNA contains a code for manufacturing individual proteins. DNA determines which proteins are made when, and which chemical reactions are controlled as a result.

For you to understand this section, you need to remember that proteins are built from chains (polypeptides) of amino acids and that there are twenty different amino acids that occur naturally. The order and number of amino acids on the chain (the primary structure) determine the physical and chemical properties of the protein.

You should also remember that DNA is made of two strands of polynucleotides wound into a double-helix structure. The two strands are held together by hydrogen bonds at their organic bases. In DNA, the base **adenine** (**A**) forms a bond with **thymine** (**T**); and **guanine** (**G**) forms a bond with **cytosine** (**C**).

> **Remember: RNA** does not contain thymine (**T**), but has the organic base **uracil** (**U**) instead. In RNA, **A** will form a bond with **U**.

DNA and the triplet code

GURU TIP

Each sequence of three bases is known as a **codon** and codes for one amino acid. Every codon is the same in every living organism (for example, CGC clodes for alanine in all organisms).

DNA carries information, which allows ribosomes to assemble specific proteins. Many of these proteins will be enzymes, controlling specific reactions. The nucleus can therefore exercise control by manufacturing specific enzymes, when required.

The order of bases along the DNA molecule forms a code for assembling and arranging amino acids along a polypeptide chain. Consider the following scenarios:

- if the DNA code was *one base codes for one amino acid*, four different codes would be possible (A, T, G or C), only producing four amino acids
- if the DNA code was *two bases code for one amino acid*, we could only get eight out of the necessary twenty amino acids (AA, AT, AG, AC, TT, TG, TC, and GC)
- to create 20 amino acids, a **triplet code** is required, to give 64 possible combinations (AAA, AAT, ATT and so on) and 64 potential amino acids.

The code is described as degenerate because there are more codes (64) than amino acids (20). Some amino acids, therefore, have more than one codon sequence describing them, for example glycine is coded by the codons CCC and CCT.

The code is non-overlapping, meaning that codons do not share their bases.

	second base				
	uracil	cytosine	adenine	guanine	
U	UUU UUC ⎤ phenylalanine UUA UUC ⎤ leucine	UCU UCC UCA UCG ⎤ serine	UAU UAC ⎤ tyrosine UAA stop UAG stop	UGU UGC ⎤ cysteine UGA stop UGG tryptophan	U C A G
C	CUU CUC CUA CUG ⎤ leucine	CCU CCC CCA CCG ⎤ Proline	CAU CAC ⎤ Histidine CAA CAG ⎤ Glutamine	CGU CGC CGA CGG ⎤ Arginine	U C A G
A	AUU AUC AUA ⎤ isoleucine AUG methionine	ACU ACC ACA ACG ⎤ threonine	AAU AAC ⎤ asparagine AAA AAG ⎤ lysine	AGU AGC ⎤ serine AGA AGG ⎤ arginine	U C A G
G	GUU GUC GUA GUG ⎤ valine	GCU GCC GCA GCG ⎤ alanine	GAU GAC ⎤ aspartic acid GAA GAG ⎤ glutamic acid	GGU GGC GGA GGG ⎤ glycine	U C A G

first base (left side label) · *third base* (right side label)

The genetic code: the base sequence in each mRNA triplet codes for a particular amino acid

| A | C | T | G | G | T | T | C | T | C | G | T | C | G | T | T | T | T | A | G | T | G | G | T | G | T | T | sugar/phosphate backbone / DNA bases |

| stop | proline | arginine | alanine | alanine | lysine | serine | proline | glutamine | codon products |

separate triplet codon

Non-overlapping codons on a short length of DNA

A **gene** (or **cistron**) is a specific section of DNA that has sufficient codons to create a particular polypeptide. The codons are always 'read' in the same direction. Some codons do not code for amino acids and therefore stop the code, these are called **stop codons**. They carry instructions to **ribosomes** (where polypeptides are assembled), marking the end of one gene and the beginning of another.

One DNA molecule contains many genes. In humans, there are about 100 000 genes formed from approximately three billion base pairs. In 1990, the US Department of Energy and National Institutes of Health set up the **Human Genome Project**. Since then, it has become a global research project with the aim of identifying all the genes and mapping the base-pairs in human DNA.

Making proteins from genes: transcription

The genetic material is held inside the nucleus and proteins are made at ribosomes on the endoplasmic reticulum. How is the DNA translated into specific proteins?

This is a two-stage process. First, a copy of a specific gene from a DNA strand is made from another nucleic acid (**ribonucleic acid – RNA**). The copy is called messenger RNA (**mRNA**) because it carries the genetic information to the site of protein manufacture (the ribosomes). The process of copying the code is known as **transcription**.

The diagram overleaf shows the stages of transcription. Try and learn them first and then apply your knowledge.

- DNA polymerase unwinds a particular section (a gene) of DNA by breaking the hydrogen bonds between the bases and exposing them. It is important that the backbone does not break at this stage (otherwise the genetic code will be distorted or even lost altogether) so there are strong and stable bonds holding it together.

- There are free nucleotides in the nucleus, containing adenine, cytosine, guanine and uracil. These are made more active by adding two extra phosphate groups.

- These bases line up against their complementary bases on the 'unzipped' DNA.

- RNA polymerase links the phosphate and sugar groups of adjacent nucleotides to form mRNA. The mRNA molecule is a mirror image of the original DNA. The two extra phosphate groups are released to activate further nucleotide molecules.

- The mRNA molecule moves away from the DNA strand, through a pore in the nuclear membrane, into the cytoplasm. Here, it attaches itself to a ribosome held in position on the surface of the rough endoplasmic reticulum.

- The hydrogen bonds between the bases of the two open strands of DNA reform. The DNA returns to its normal double-helix structure.

The total set of genes in a cell is known as the **genome**. Since all cells (with certain exceptions, for example red blood cells and gametes) will contain exactly the same set of genes, the genome represents the total set of genes in a particular organism.

GURU TIP
You can find out about the status of the project, and much more, from the Internet.

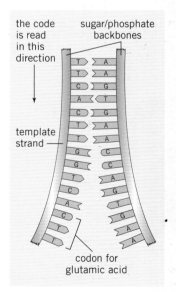

the code is read in this direction

sugar/phosphate backbones

template strand

codon for glutamic acid

DNA codons don't overlap

PS3, LP3

Proteins: transcription and translation

Making proteins from genes: transcription

Once you've learnt the stages of transcription, try and follow the diagram. It's important to understand this before you start on the next step in protein synthesis.

GURU WEBSITE

If you're finding AS Biology difficult, or have any questions or comments, leave a message or visit the message board on the AS Guru™ Biology website and see what advice other students and even teachers have to offer.

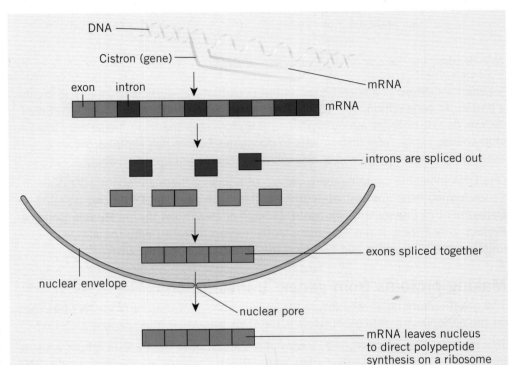

DNA

Cistron (gene)

mRNA

exon   intron

mRNA

introns are spliced out

exons spliced together

nuclear envelope

nuclear pore

Transcription is the first stage in protein sysnthesis

mRNA leaves nucleus to direct polypeptide synthesis on a ribosome

Translation

The genetic code needs to be copied to a mRNA molecule. Then the information is **translated** to manufacture protein molecules at the ribosomes in the cytoplasm. This is achieved by using one of 20 specific **transfer RNAs (tRNA)**.

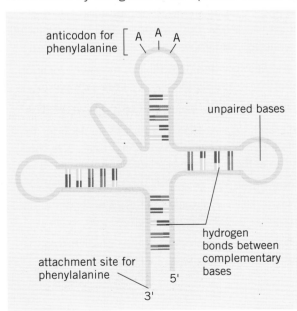

anticodon for phenylalanine   A A A

unpaired bases

hydrogen bonds between complementary bases

attachment site for phenylalanine

5'

3'

tRNA transports amino acids to the ribosome

There is a triplet of nucleotides called the **anticodon**, which are exposed at one end. At the other end there is an attachment site for a specific amino acid. This corresponds to the anticodon and the codon on the mRNA.

Each tRNA picks up a specific amino acid and transports it to the ribosome. By matching the anticodon to the mRNA codon, the amino acids are assembled in the correct order. Amino acids need activating by ATP and a specific enzyme before joining a tRNA, ensuring that the correct amino acid and tRNA combine.

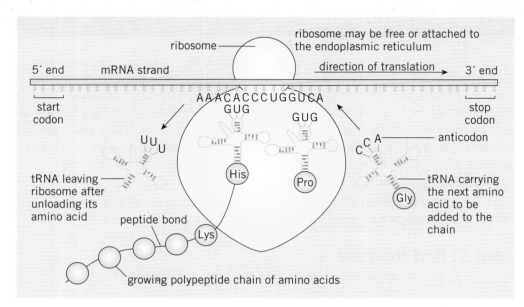

Translation of the genetic code

Occasionally, there are **mutations** (spontaneous changes) to DNA caused by radiation exposure, carcinogens or faulty duplication.

Mutagens can cause spontaneous changes to the genetic code

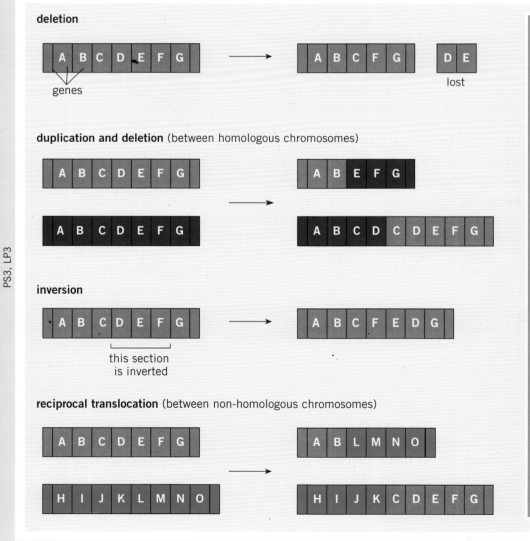

- **Substitution**: one base in a codon is altered.

- **Inversion**: two bases in a codon swap over.

- **Deletion**: one codon is missing, so the polypeptide has one less amino acid.

- **Addition**: an extra codon appears.

- **Translocation**: a section of DNA breaks off and attaches to another.

- **Duplication**: a DNA section duplicates, repeating the gene.

- **Non-disjunction**: replication of one or more strand of DNA, producing extra copies. Alternatively, one or more strands of DNA may be missing.

PS3, LP3

Genetic engineering

AQA A
AQA B
OCR

Genes control cells and therefore whole organisms. Genes that code for something that helps an individual to survive and breed get passed on to the next generation. Humans have manipulated this process for centuries by selectively breeding crops and animals, choosing the individuals with genes that code for resisting a disease or making more milk. With luck, breeding from such individuals will, over time, get you what you want. Now, genetic engineering can specifically change an organism's **genome**. Genes can be taken from one organism and inserted into an entirely different species, creating a **transgenic** organism.

Here is how it works, step by step, taking human insulin as our example. This is harder than it sounds and there are three approaches:

Step 1: find the gene

- If you know part of the base sequence of the gene you're after, you can use a **probe** (a single-stranded piece of DNA with a complementary base sequence to the one you're looking for) to pair with your gene. Label it with a fluorescent dye or radioactive isotope, to identify it.
- Find cells that are making insulin (β-cells in the pancreas) as the insulin-coding gene will be switched on here and will be producing mRNA. Isolate this mRNA, label it and use it as a gene probe. Or use an enzyme called **reverse transcriptase** to make a DNA copy from it. This is called **complementary DNA (cDNA)** and is an exact copy of the original DNA, without any sections that don't actually express themselves (introns). For insulin, this was the approach that worked.
- Work backwards from the protein, working out the amino acid sequence and then making a piece of cDNA that codes for the protein.

Step 2: transfer the gene

Having found the piece of DNA, containing the insulin gene, you need molecular 'scissors' and 'glue' to cut it out and stick it into the genome of another organism.

Restriction endonucleases act as scissors. They seek out a **recognition site** (specific base sequence) and snip the DNA at this point. They make staggered cuts, leaving **sticky ends**. These combine with complementary sticky ends, so lengths of DNA from one organism can be spliced into the DNA of another. The join is made permanent using **DNA ligase**. 'Ligate' means 'to join' – this enzyme acts as glue.

> An example of a restriction enzyme is EcoRI. It originates in *E. coli* bacterium and works on the following recognition site, cutting both strands:
>
> G | A A T T C
> C T T A A | G

Bacteria contain rings of DNA called **plasmids**, separate from their own chromosomes. This is handy – you can remove a plasmid from a bacterium and cut it with the same restriction endonuclease you used on the human DNA so that it leaves sticky ends that match the DNA fragment you want to splice in. Mix them together with some DNA ligase and the plasmid might take up the human gene. Put the altered plasmid back into a bacterium, and you have your **genetically engineered transgenic bacterium**. As the plasmid has been used as a carrier to get the new gene into the bacterium, it is called a **vector**.

> The Human Genome Project has worked out the base sequence for the entire genome. Unfortunately, the researchers don't know what most of these genes actually code for, so there is still a long way to go.

Step 3: clone the gene

To make useful amounts of insulin, there need to be many copies of **recombinant DNA**. The transgenic bacteria are put into a **fermenter** and provided with the ideal conditions for growth. The bacteria multiply, dividing into two clones about every half hour. Inside each bacterium, the recombinant plasmids replicate, making many clones. You soon have millions of identical copies of the gene for insulin.

Step 4: collect the insulin

The insulin gene in the cloned plasmids expresses itself. The bacteria produce large amounts of human insulin ready to be extracted, purified and used to treat diabetes.

Insulin is a protein belonging to a group of chemical messengers called hormones. It helps control the amount of sugar in your blood. People who are diabetic can't make insulin for themselves, so they have to inject it. Insulin extracted from animals works but it is not identical to human insulin and there is always a risk of infection. Genetic engineering produces pure human insulin, in high volumes and cheaply.

Genetic engineering, DNA technology and recombinant DNA technology amount to the same thing – moving genes from one organism into another. Biotechnology is the use of micro-organisms (such as bacteria and yeast) to make useful substances.

Restriction endonucleases are made by bacteria to beat off invading viruses, called bacteriophages. They restrict the damage by snipping inside the nucleic acids of the bacteriophage DNA. The snipped fragments can't make fresh copies of the virus.

Getting a bacterium to accept the genetically altered plasmid isn't easy. Soaking the bacteria in ice-cold calcium chloride then incubating them for 2 minutes at 42°C sometimes does the trick, although no-one's sure why.

Knowing exactly which bacteria are recombinant is another problem. Two genes can be inserted into the plasmid: the one you want and one that makes the bacterium easy to spot – a **genetic marker**. This could make the recombinant bacteria resistant to a particular antibiotic. Press a sterile pad onto the master colony to pick up the bacteria and then transfer them to a plate with antibiotic added to the growth medium (**replica plating**). Only colonies of recombinant bacteria will survive.

The same techniques have been used to get micro-organisms to make antibiotics and enzymes to use in washing powders, as well as hormones, such as like insulin.

GURU WEBSITE
All the core skills for AS Biology are explained on the website. So you'll know exactly what's expected of you.

Step by step process in gene engineering

1 isolate the gene and cut out a fragment, containing the gene, using a restriction enzyme

human cell — chromosome — DNA strand — restriction enzyme — gene coding for useful product — sticky ends

2 isolate plasmid from a bacterial cell, using the same restriction enzyme

bacterial chromosome
bacterial cell
sticky ends

3 splice the human DNA to the plasmid, using DNA ligase to join the sticky ends

bacterial plasmid
human gene
joint between human and bacterial DNA made by ligase enzymes

4 ensure the bacterium takes up the recombinant plasmid and allow it to multiply so that the human gene or product of the gene can be used

C3, IT3, LP3 and W03

Medical uses of gene technology

AQA B
WJEC

Making human factor VIII (pronounced 'factor eight')

Factor VIII is a blood protein that is needed to make blood clot. People who suffer from the inherited disease **haemophilia** can't make their own and need to give themselves regular injections of factor VIII. It used to be extracted from donated blood but this wasn't ideal because of the infection risk. One injection could contain factor VIII from several donors, any of whom could be carrying an infectious disease. Indeed, many haemophiliacs became infected with HIV in this way. Now, the gene for making human factor VIII has been inserted into hamster kidney cells and ovary cells. The transgenic cells are cultured in fermenters, producing a steady supply of safe factor VIII to be extracted and purified.

Transgenic sheep

A new approach is to insert human genes into the genome of animals that make milk, combined with a promoter sequence that switches the gene on in their mammary glands. Human alpha-1-antitrypsin (AAT) is a protein used to treat cystic fibrosis and the lung disease, emphysema. By injecting plasmids containing the AAT gene into fertilised egg cells, a sheep (called Tracy) was born in 1993 who made AAT in her milk. She even passed this ability on to one of her lambs. There is the prospect soon of having whole flocks of transgenic sheep producing valuable human proteins such as AAT and factor VIII in their milk.

Gene therapy to treat cystic fibrosis

Instead of making the missing protein and giving it to the patient, it may sometimes be possible to clone healthy genes and put them into cells in the patient's body, where they can take over from the defective genes that are causing the illness. This is the idea behind **gene therapy**. There are some obvious problems – getting the replacement genes into enough of the right cells (and expressing themselves) to make a difference, for a start. Also, cells are constantly dying and being replaced by new cells, so repeat treatments are needed.

Cystic fibrosis is a genetic disease. Sufferers have inherited two copies (one from each parent) of a faulty gene. Normally, this gene would make a protein called **cystic fibrosis transmembrane regulator** (CFTR), but because of a mutation in the DNA, just one of the 1480 amino acids in this protein molecule is missing. This is enough to change the protein's shape and stop it working. CFTR is a channel protein. It has an important job transporting chloride ions out of cells through the cell membrane. The defective CFTR protein lets these ions build up inside cells, causing them to retain water. It affects all the cells in the body, but is a particular problem in the lungs and intestines. Here, the normally runny mucus becomes thick and sticky because of water being held back in the lining cells. This makes breathing difficult and provides a breeding ground for harmful bacteria. The duct carrying digestive enzymes from the pancreas to the gut also gets blocked, interfering with normal digestion.

A new gene therapy involves inserting cloned, healthy CFTR genes into plasmids and then wrapping the recombinant plasmids up in tiny lipid droplets called **liposomes**. Using something like an asthma sufferer's inhaler, a fine aerosol spray of liposomes can be inhaled into the lungs, where they fuse with the target cells' phospholipid membranes and smuggle the working genes inside.

GURU TIP

Check the specifications for your exam body, to see which particular examples of gene technology are mentioned. Learn about those examples in extra detail.

Biomedical companies spend millions on searching for new treatments. They need to get a return on their investment, to fund further research and pay shareholders. One way of doing this is to take out patents, so that any future drugs or therapy based on the patented genes would belong to the company that developed it.

Some people say that the human genome isn't for sale.

What do you think?

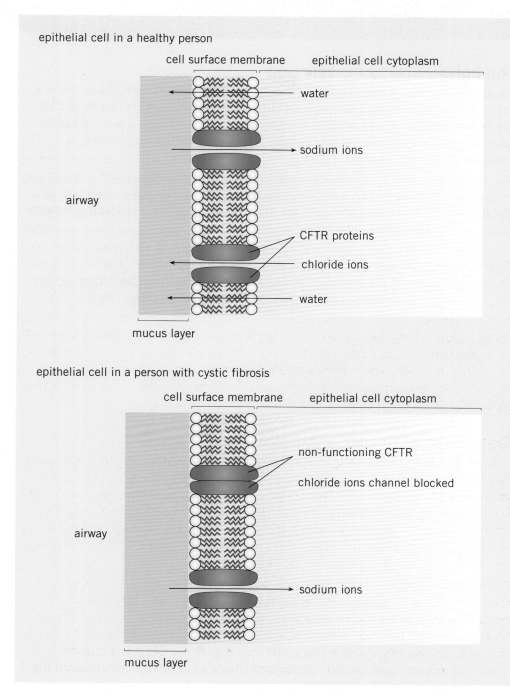

epithelial cell in a healthy person

cell surface membrane | epithelial cell cytoplasm

- water
- sodium ions
- CFTR proteins
- chloride ions
- water

airway

mucus layer

epithelial cell in a person with cystic fibrosis

cell surface membrane | epithelial cell cytoplasm

- non-functioning CFTR
- chloride ions channel blocked
- sodium ions

airway

mucus layer

Transport of ions by CFTR

Viruses are experts at inserting their DNA into cells. This is how they reproduce, hijacking the host cell's biochemical machinery to make copies of themselves. An alternative approach to gene therapy exploits this by getting **adenoviruses**, which normally cause colds, to act as **vectors**, carrying healthy genes into the target cells. First, the viruses are rendered harmless by disabling the genes they need to copy themselves. Then, the modified adenoviruses are cultured with recombinant plasmids containing healthy CFTR genes, which become part of the virus genome. Sprayed into the lungs, the adenoviruses 'infect' the cells lining the patient's lungs. But instead of catching a cold, the CFTR protein is made as normal (at least for a while). Unfortunately, the patient's own immune system may make antibodies to destroy these 'invading' viruses, making the patient resistant to the treatment.

PCR and DNA fingerprinting

→ AQA A
→ AQA B

DNA polymerase occurs naturally in cells. It catalyses the replication of DNA, ready for cell division. The DNA polymerase used in PCR comes from bacteria that live in hot springs, so it is thermostable – it can work at high temperatures without denaturing.

GURU TIP
Geneticists often call a standard procedure a *protocol*. This just refers to the series of steps to go through when cloning a particular gene, for example.

Comparing DNA fingerprints

The polymerase chain reaction

The polymerase chain reaction (PCR) is a revolutionary technique that can make a billion copies from a single strand of DNA in a few hours. Useful amounts of DNA can be **amplified** from tiny quantities extracted from bits of hair, blood, semen or even sweat found at a crime scene, ready for analysis and comparison with a suspect's DNA.

The first stage is to mix the original DNA sample with DNA polymerase and a supply of nucleotides. Next, add some short pieces of DNA called **primers**, which act as signals to the enzyme saying 'start copying here'. Heat until the two strands of the DNA helix unravel, revealing their base sequences. The enzyme uses the nucleotides to assemble a fresh strand of DNA along each of the originals with a matching complementary base sequence, just as in normal DNA replication. The result is two perfect copies of the DNA molecule you started with. Repeating the process yields four copies, then eight, then sixteen, and so on – this is the 'chain reaction' bit.

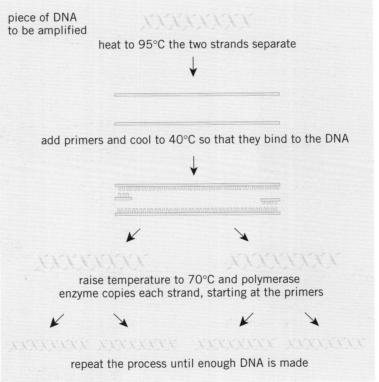

piece of DNA to be amplified

heat to 95°C the two strands separate

add primers and cool to 40°C so that they bind to the DNA

raise temperature to 70°C and polymerase enzyme copies each strand, starting at the primers

repeat the process until enough DNA is made

A single PCR cycle

DNA fingerprinting

Having got enough DNA, it is snipped up into short lengths using restriction endonucleases. The target base sequences, where the enzymes cut, are irregular distances apart so a mixture of fragments of different length is left. The relative amounts of fragments, vary between individuals, even from the same family. The next trick is to spread the fragments out and make them visible, ready for comparison with the suspect's DNA.

Gel electrophoresis is used to sort out the DNA fragments. The solution containing the scissored fragments of DNA from the crime scene is put into a well at one end of a long tray of agar gel. The suspect's DNA fragments are placed in a well alongside. A buffer solution that conducts electricity is poured on top. Electrodes at either end then apply a voltage. The phosphate groups on the DNA fragments give them a negative charge, so they are attracted through the gel towards the positive electrode. The smaller fragments move faster, so the fragments spread out along the gel forming invisible bands sorted in order of length.

To see where the fragments have ended up, a nitro-cellulose sheet is pressed against the gel. It soaks up the DNA fragments like blotting paper. A radioactive DNA probe is added, which binds to specific DNA sequences. Holding the sheet against a piece of unexposed film lets the radioactivity fog the film wherever the DNA probe has accumulated. The developed film is the finished DNA 'fingerprint'.

A match between the pattern of strips from the crime scene DNA and from the suspect's DNA is very unlikely to happen by chance. Any difference in the pattern, however, proves the suspect's innocence beyond reasonable doubt.

GURU TIP
You should be able to show that you are aware of the moral and ethical issues involved in the gene technologies described here.

Enzyme digestion cuts DNA into fragments

Electrophoresis sorts DNA fragments

DNA extracted and purified

Blood sample

DNA transfer from gel to nylon membrane

Results show DNA bands

Film exposure detects position of probes

Hybridisation probes bind to certain DNA fragments

Analysing DNA samples

Gel electrophoresis

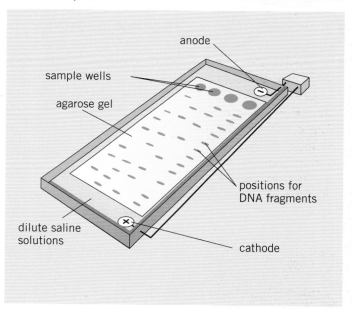

anode

sample wells

agarose gel

positions for DNA fragments

dilute saline solutions

cathode

Only about 2% of the human genome actually codes for real, useful proteins. The rest is 'junk' DNA' that never gets used. In amongst the junk there are great long sections of a few bases repeated over and over again. The length and number of these **tandem repeats** varies from one person to another – an ideal target for genetic fingerprinting.

Gel electrophoresis of DNA labelled with a flourescent dye

Summary

DNA contains genes, which code for proteins. Triplets of bases, called codons, code for individual amino acids. When a protein is built, the DNA unzips and the code is read. By assembling amino acids in the correct order, any protein can be made. Several mRNA copies of a gene can be made, so lots of protein molecules can be made quickly. The two stages in protein synthesis are transcription (copying the code to mRNA) and translation (using tRNA to string amino acids together).

Polypeptides can be anything from structural proteins to the enzymes that control a cell's biochemical machinery. By directing the production of proteins, DNA controls every aspect of the cell's structure and function.

Mutations are changes in the genetic code. They are usually harmful but can occasionally cause an improvement in the protein. Mutations are important in evolution, as they are the only way that completely new genes can appear.

Genetic engineering alters an organism's genome so that it makes or does something we want. One way is to transfer a gene from another organism. This is possible because the genetic code is universal.

DNA fingerprinting relies on DNA being specific to every individual.

Practice questions

1 The diagram shows part of a messenger RNA (mRNA) molecule:

U A A U A C C G A C C U U A C C C U

  a Write out the sequence, adding the cDNA sequence from which it was transcribed.

  b How many codons are shown on this section of mRNA?

  c Describe how tRNA molecules could translate this sequence into a polypeptide.

2 Describe how deletions, substitutions and inversions cause mutations in the genetic code and the effect it has on the polypeptide chain that is produced.

3 Describe how recombinant DNA technology (genetic engineering) was used to produce transgenic bacteria that make human insulin. Write your answer in continuous prose (not notes), taking between one and two sides of A4.

Key skills

Can you clearly describe the way in which a nucleotide sequence codes for an amino acid sequence in a polypeptide? If you are able to do this, you can present the information in various formats, in order to satisfy all of PS3 and LP3 key skills. Can you build a three dimensional model of the translation and transcription process? Go about it the right way and you can get credit for all of PS3.

Try and find up-to-date research on the synthesis of human factor VIII and describe the process simply and clearly. If you demonstrate that you understand the various techniques used and can present your findings appropriately, you can easily fulfill all of C3, IT3, WO3 and LP3 key skills.

Go through the checklist in the book's introduction to see what you need to do.

Reproduction

You already know that reproduction is a key characteristic of life. Many biologists would even say that everything about an organism is adapted in order to improve the chances of reproducing successfully and passing on its genes. According to this 'selfish gene' theory, evolution generates increasingly sophisticated bodies as vehicles to pass on their genes to the next generation. The genes are in the driving seat and even our complex human bodies are ultimately packages for transferring genes from one generation to the next. You only need to look at the lengths that organisms go to in order to reproduce to realise that it is a fundamental part of life.

The nucleus of every normal cell in your body contains 23 pairs of chromosomes. The DNA in these chromosomes holds all the genetic instructions to make a complete human body. All the billions of cells of your body are the direct descendants of a single fertilised cell. How each cell 'knows' what sort of cell it's supposed to be is a bit of a mystery. How do all the right genes get switched on or off in the right places at the right times? However this works, cells go through a life cycle. First the DNA copies itself, then the whole cell divides to produce two genetically identical copies (clones). This is how organisms grow, repair themselves and reproduce asexually. So where does sex come into it?

It might surprise you to know that biologists are still arguing about how sex evolved, and why so many organisms have two sexes – slime moulds have thirteen! It is clear that sexual reproduction is brilliant at producing variety. Out of five billion people, nobody is exactly the same, unless you happen to be an identical twin. This is down to the random shuffling of genes that occurs when egg and sperm are made by meiosis. Throw in the lottery of which sperm fertilised which egg, add a dash of random mutation and it's no surprise that we are all genetically unique. Sexual reproduction may seem like a lot of trouble but the resulting variation gives it a massive advantage over asexual reproduction. Variation allows fresh solutions to the problems faced by a species which produces individuals that are better adapted than their ancestors and, given enough time, whole new species can appear.

The cell cycle – the life story of a cell

Normal body cells have 23 pairs of chromosomes, that's 22 matching homologous pairs and one pair of sex chromosomes (XX in girls, XY in boys). These are called diploid cells (2n for short). Sperm and eggs only have half the normal number of chromosome, one from each pair, so they're called haploid cells (n for short).

Clones are everywhere!

Clones are **genetically identical** cells or organisms. Human identical twins are clones. Any organisms that reproduce asexually (without sex) are also making clones of themselves. Blackberry plants for example, make branches that put down roots that grow into clones of the parent plant. Crop plants reproduce asexually, which has implications for the farmer. This is a bonus in some respects, as all the plants have consistent characteristics and quality but, less helpfully, they also have the same susceptibilities to pests and diseases. When plants reproduce asexually, they are making natural clones. This process is called **vegetative propagation**.

Making genetically identical offspring calls for a kind of cell division called **mitosis**. The result of mitotic division is two cells, each containing an exact copy of the original (mother) cell's genetic code. Mitosis is happening in your body right now.

> **Remember**: when you grow your cells don't get bigger – you just make more of them. Also, your cells are wearing out all the time and need replacing. The stuff that fills up your vacuum cleaner is mainly dead skin cells from you and your family.

Let's think about a single cell, and the stages it has to go through during its life. The scientific name for this is the **cell cycle**:

GURU TIP

Mitosis occurs whenever there is a need for new, identical copies of a cell: for **growth**, **repair** and **replacement** as well as **asexual reproduction**.

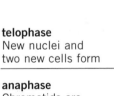

growth
Cell produces proteins, growth and specialisation occur

interphase

replication
DNA molecules are copied

prophase
Chromosomes coil and shorten, revealing two chromatids joined at centromere

metaphase
Chromatids attach to spindle at equator of cell

anaphase
Chromatids are pulled to poles of cell

telophase
New nuclei and two new cells form

Mitotic cell cycle

KEY SKILLS
C3, PS3, LP3

Interphase

This is the largest part of the cell cycle, when the cell grows and specialises. The DNA molecules stretch out, synthesising proteins and governing the cell's activities. When the cell is preparing to divide, the DNA replicates itself, making two identical copies of each double helix. The cell is ready for mitosis to begin.

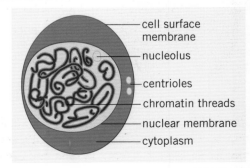

- cell surface membrane
- nucleolus
- centrioles
- chromatin threads
- nuclear membrane
- cytoplasm

Prophase

The chromosomes coil up and shorten. They can now be seen, under a light microscope, as double strands or chromatids. The 'sister' chromatids are firmly attached to each other at the centromere. Each pair contains two identical copies of DNA.

The **centrioles** (bundles of tiny protein fibres) move towards opposite ends of the cell. These start to form a web of microtubules across the cell, called the spindle.

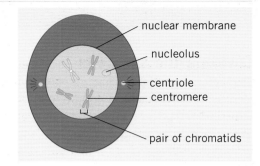

- nuclear membrane
- nucleolus
- centriole
- centromere
- pair of chromatids

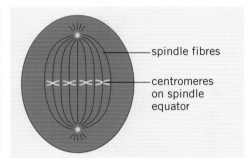

- spindle fibres
- centromeres on spindle equator

Metaphase

The membrane around the nucleus breaks up. The centromeres attach themselves to the spindle, which is now fully developed, and line up along the middle of the cell.

Anaphase

The pairs of sister chromatids split apart and move along the spindle towards opposite ends of the cell.

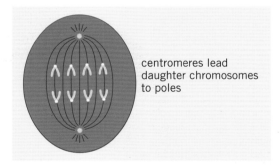

centromeres lead daughter chromosomes to poles

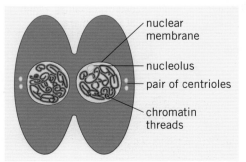

- nuclear membrane
- nucleolus
- pair of centrioles
- chromatin threads

Telophase

The chromatids (exact copies of the original chromosomes) group together at opposite ends of the cell and start to uncoil so they can no longer be seen. Nuclear membranes form around them, the cytoplasm divides and a new cell membrane forms, separating the two daughter cells (in plant cells, this is when they make a new cell wall).

One cell has divided to make two identical cells. Mitosis is finished. The new cells now specialise to form any type of cell, from liver cells in the liver, to palisade mesophyll cells in a leaf, and the cell cycle starts all over again.

Meiosis

KEY SKILLS C3, PS3, LP3

When two gametes join at fertilisation, the homologous (matching) chromosomes pair-up, making a **zygote** containing the full, diploid set of chromosomes.

Dividing the sex cells

Gametes (sex cells) are haploid cells, with half the number of chromosomes found in diploid cells. A special kind of cell division takes place when gametes are made, called **meiosis**.

- The cells that are produced at the end of a meiotic cycle are haploid (they only contain half the number of chromosomes of a normal cell). If you think about it, this has to be the case, otherwise the number of chromosomes would double with each generation, every time an egg and a sperm joined.
- The cells produced by meiosis are genetically different from the original cell and, indeed, from each other. This shuffling of genes is one of the reasons that organisms that reproduce sexually show so much variation between individuals.

Because it reduces the number of chromosomes, meiosis is sometimes called **reduction division**. It reduces the diploid number of chromosomes (2n) to the haploid number (n). In humans, an egg or sperm has 23 single chromosomes, one from each of the 23 pairs of chromosomes in a normal cell.

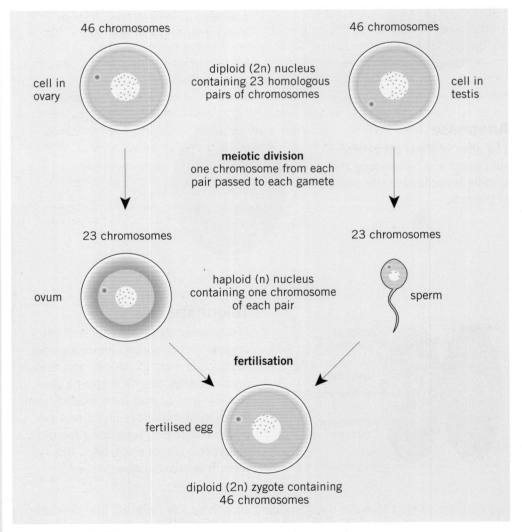

Meiosis and fertilisation in humans

Meiosis is quite complicated but there are a few important points. Like in mitosis, DNA replication happens in interphase. Unlike mitosis however, there are then two rounds of nuclear division:

• division 1 separates the homologous pairs of chromosomes
• division 2 separates the pairs of chromatids that make up each chromosome.

The end result is four haploid nuclei. Meiosis happens in animals when gametes are made (gametogenesis) and when plants make spores (sporogenesis).

Something special happens during **prophase I**. Unlike mitosis (where the chromosomes of each homologous pair do their own thing), in meiosis they pair-up.

Remember: DNA has copied itself already, so each chromosome consists of two identical sister chromatids.

Here's the clever bit: non-sister chromatids undergo a process called **crossing over**. This is when the chromatids form links to one another. The points at which they join are called **chiasmata** (pronounced 'ky-as-mata' from the Greek for 'cross arrangement', singular: chiasma). The chromatids break and rejoin at the chiasmata, swapping sections of genetic code with one another. This provides new combinations of genes, mixing up genetic information from the paternal and maternal chromosomes and adding to the genetic variation in the resulting nuclei. It is completely random which chromosome of each pair ends up in each nucleus. This **independent assortment** of chromosomes adds even more variation.

Given that it's down to blind chance, which of billions of genetically different sperm will happen to fertilise a particular egg, it's really not such a surprise that sexual reproduction produces genetically unique individuals.

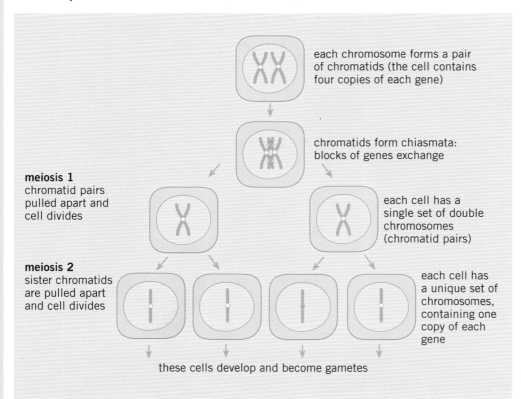

meiosis 1
chromatid pairs pulled apart and cell divides

meiosis 2
sister chromatids are pulled apart and cell divides

each chromosome forms a pair of chromatids (the cell contains four copies of each gene)

chromatids form chiasmata: blocks of genes exchange

each cell has a single set of double chromosomes (chromatid pairs)

each cell has a unique set of chromosomes, containing one copy of each gene

these cells develop and become gametes

Behaviour of chromosomes during the stages of meiosis.

GURU TIP
Don't get too hung up on the details of meiosis. Make sure that you understand what's happening to the chromosomes, and how this leads to haploid cells, which show genetic variation – that's the whole point of meiosis.

C3, PS3, LP3

Reproduction in flowering plants

The problem with clones

Making new cells by mitosis allows plants and some animals, to reproduce asexually by making genetically identical clones of themselves. Whilst this is a quick and easy way to reproduce, it has one big drawback.

During the Irish potato famine, millions of people starved because the entire potato crop was wiped out by a fungus. All these plants were clones. If there had been some genetic variation, maybe some of the plants would have been able to fight the disease. These could then have been cultivated into a successful new variety. This is where sex comes in. Because meiosis reshuffles genes and sex allows genes from two parents to blend together, sexual reproduction generates genetically unique individuals. Some of these, by luck of genetic variation, will be better at coping with their environment than their parents or siblings. This variation is the basis of evolution through natural selection and is an important biological theory.

Why do we need flowers for reproduction?

Flowers facilitate sexual reproduction. Because plants are unable to move, they need to get the male gametes (packaged in pollen grains) to the female parts of other flowers (pollination). They can do this in two ways:

- throw out loads of pollen for the wind to blow about, in the hope that some will land on another flower of the same species (wind pollination)
- bribe insects with nectar to carry pollen between flowers (insect pollination).

Flowers attract insects, which pollinate them

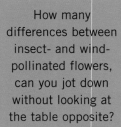

GURU TIP
Don't get confused between pollination and fertilisation. **Pollination** is the movement of pollen from anther to stigma. **Fertilisation** is the fusion (joining) of a male gamete with a female gamete to produce a zygote.

How many differences between insect- and wind-pollinated flowers, can you jot down without looking at the table opposite?

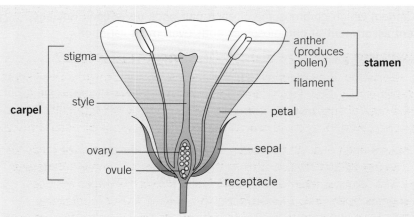

Section through an insect-pollinated dicotyledonous flower

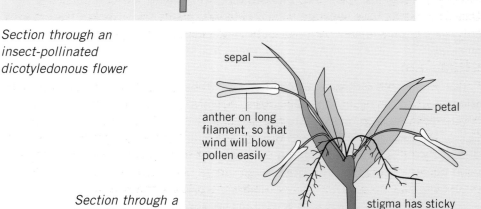

Section through a wind-pollinated grass flower

> **Stamens** – male organs that consist of anthers held up by filaments. The anthers make pollen grains, containing a male gamete with a haploid set of chromosomes produced by meiosis. They are designed to dump pollen onto visiting insects.
> **Carpels** – female organs that consist of a stigma, style and an ovary. The ovary holds the haploid ovules. The stigma is sticky and picks up pollen from insects.
> **Petals** – big, colourful and often scented, advertising the nectar in the flower to the right species of insect for pollination.
> **Sepals** – protective casing around the flower bud, before it blossoms.

Comparison of wind- and insect-pollinated plants

Insect pollinated plants	Wind pollinated plants
Big, colourful petals to attract insects.	Small, green petals if any at all.
Often scented to attract the appropriate insect. Not always pleasant – *Fritillaria* flowers smell of rotting flesh!	Never scented.
Stamens and stigma inside flower, arranged so insects brush against the pollen, transferring it to the stigma of another flower of the same species.	Stamens dangle outside flower so pollen is blown away by wind. Large, feathery stigmas have long styles, dangling outside the flower to catch pollen.
Nectaries make an energy-rich nectar to attract insects.	No nectaries.
Small amount of pollen made, as it is carried straight to a recipient stigma with much less wastage. Grains tend to be rough so that they stick to insect bodies.	Huge amount of pollen made to improve chances of landing on a stigma. Pollen grains are smooth and light so the wind carries them far and wide.

How do flowering plants avoid pollinating themselves?

Self-pollination defeats the point of reproducing with gametes – you might as well stick to asexual vegetative propagation. Fresh combinations of genes have the potential for new and improved ways of surviving. Most flowers are hermaphrodites with both male and female organ, so how do they avoid self-pollination? The answer is all about timing.

- **Protandry** – anthers ripen first. All the pollen is made before the ovules are ready for fertilisation. *Rosebay Willowherb* does this so that only flowers on other *Willowherbs* are pollinated. The favour is returned when the first *Willowherb's* ovules are ripe.
- **Protogyny** – stigmas ripen first. *Bluebells* use this reverse form of protandry.
- **Dioecious** plants have separate male and female plants. *Holly* makes flowers with only male or only female parts, so there is no chance of self-pollination.

Fertilisation

- Pollen grains land on the stigma and a pollen tube grows down, through the style and ovary, into an ovule. A **tube nucleus,** near the tip, makes digesting proteins.
- A **generative nucleus** divides into two nuclei and these follow the tube nucleus.
- One of the two male nuclei fuses with an egg cell, to make a diploid zygote. This divides by mitosis to make an embryo. The other male nucleus fuses with the two **polar nuclei**, making a triploid cell (three sets of chromosomes), which divides to produce a food store for the embryo called the **primary endosperm**.

There are two sets of nuclear fusion: one to form the zygote and one to form the primary endosperm – **double fertilisation**. The whole thing now grows into a seed.

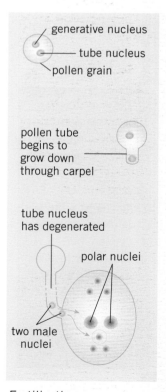

Fertilisation

C3, N3, IT3, WO3, LP3, PS3

Reproduction in humans: male

The male reproductive system

The job of this organ system is to produce male gametes (sperm), together with the fluid in which they swim, and deliver them into the vagina.

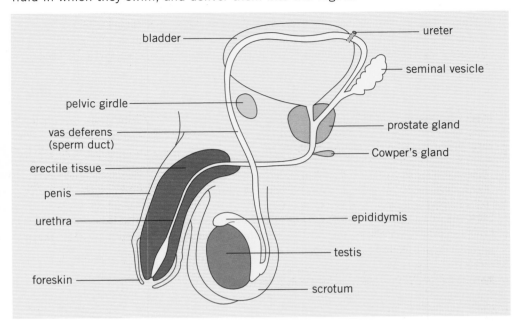

The male reproductive system

GURU TIP
You'll remember much of this from GCSE. Now try and learn the extra detail. Try covering up the labels and see if you can recall them all from memory.

There is only one tube in the penis that carries urine out of the body from the bladder, when limp. During an erection, it connects with the vas deferens, carrying semen instead. A vasectomy involves cutting the vas deferens above the testes so that semen is still ejaculated, but contains no sperm.

You should learn a little more detail than you needed for GCSE, in particular:

- the **epididymis** is where sperm are stored
- the **vas deferens** is the sperm tube
- the **seminal vesicles** add four things to semen: fructose (provides sperm with energy), mucus (for lubrication), protein (gives semen the right consistency) and **prostoglandins** (stimulate peristalsis in the female system)
- the **prostate** adds alkalis to neutralise acid in the vagina, and a clotting agent that acts on the protein to give semen a jelly-like consistency
- the **Cowper's gland** adds a fluid to clean the urethra just before ejaculation
- the **ureters** carry urine from the kidneys and join up with the urethra, which carries either urine or semen through the penis and out of the body
- sperm production (spermatogenesis) is most efficient at about 35°C, just below body temperature (37°C). That's why the testes hang, rather vulnerably, outside the body in a sack of skin called the scrotum.

An erection happens when the blood vessels leading to the penis dilate (widen) and allow the spongy erectile tissue to fill up with blood. It is the pressure of the blood that makes the penis stiff – there are no muscles involved. During ejaculation, strong waves of peristalsis push sperm along the vas deferens, past glands that add fluids to make semen, along the urethra and out of the penis. A typical ejaculation produces about 5cm³ of semen, containing 50–200 million sperm.

Spermatogenesis

Sperm are made at a rate of around 1000 per second, from puberty well into old age. Each testis (plural: *testes*) consists of lobules of **seminiferous tubules**, where sperm are made. The testes also make the male sex hormone **testosterone**.

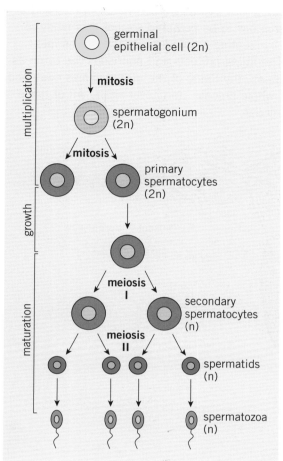

Stages of spermatogenesis

There are three stages of spermatogenesis:

- multiplication – germinal epithelium cells mitotically divide to make lots of **spermatogonia**

- growth – the spermatogonia (diploid), grow into **primary spermatocytes**

- maturation – the primary spermatocytes meiotically divide. The first division produces haploid **secondary spermatocytes**. These divide once more, to produce **spermatids**, which mature into spermatozoa (sperm).

While they are developing, the sperm cells are nourished by **Sertoli cells** (nurse cells). The fully developed sperm are stored in the **epididymis** (a long, coiled-up tube behind the testis).

Spermatogenesis is a complicated process and errors are quite common. Many sperm are deformed, possibly with two tails or just whizzing around in circles. It takes a while, too. By the time they are ejaculated, the sperm are about three months old.

Scanning electron micrograph of sectioned seminiferous tubules

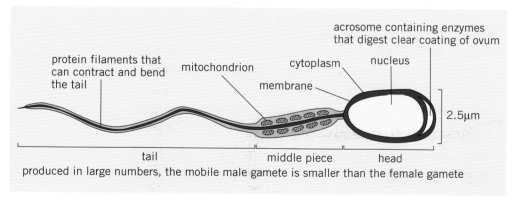

produced in large numbers, the mobile male gamete is smaller than the female gamete

Sperm cells are just 2.5µm across

Each sperm cell has a head, a middle piece and a tail. The head contains the nucleus with its 23 chromosomes carrying the all-important genetic information in the DNA. It is capped by the **acrosome**, which contains the protease enzymes needed to penetrate an egg. The middle piece is stuffed full of **mitochondria** – the power-house organelles where respiration happens. Respiration releases the energy needed to make the tail thrash from side to side.

The female reproductive system

This organ system has a more complicated job. Ova (eggs) are matured, released and transported to the uterus. The uterus has to be prepared so that it is ready to accept a fertilised ovum. This is done by creating a lining inside the uterus into which the fertilised egg can implant. The lining protects and nourishes the ovum as it develops. If fertilisation does not happen, the lining of the uterus comes away and passes out through the vagina and the whole process starts over again. All of these events need to be co-ordinated by hormones.

GURU TIP

Remember that ova, like sperms, are produced by meiosis. They are haploid (n) cells, containing just 23 chromosomes

A human egg cell is the largest cell in the body (over 0.1mm across). You can just see an ovum without using a microscope.

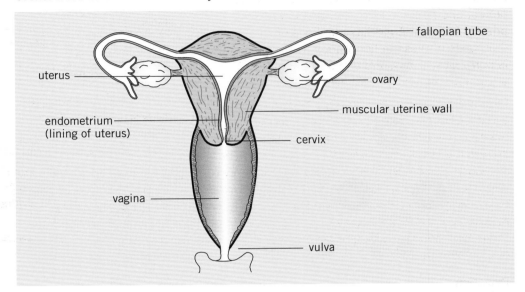

The female reproductive system

The menstrual cycle stops at the menopause, when the ovaries stop working, usually between the ages of 45 and 55. The reduced levels of oestrogen and progesterone that cause the menopause also have several unpleasant side effects. These can be reversed by hormone replacement therapy (HRT).

Try to learn these details:

- ovaries develop the ova and make the hormones oestrogen and progesterone
- an oviduct (fallopian tube) connects each ovary to the uterus (or womb)
- the oviducts are lined with ciliated epithelia, which move the egg along with a wafting motion
- glandular cells secrete mucus, which stops the ovum from drying out
- fertilisation occurs in the oviduct
- the uterus has a thick wall of muscle and is lined by the endometrium
- the endometrium is a mucus membrane with a good supply of blood capillaries to supply the placenta during pregnancy
- the top layer of the endometrium is lost and replaced each menstrual cycle
- the cervix is a narrow muscular channel at the bottom of the uterus, connecting to the vagina, which is a muscular tube leading out of the body.

During sex, the vaginal muscles relax and glands in the lining produce mucus for lubrication. When a baby is being born, the cervix opens up (dilates) to about 10cm.

Oogenesis (making ova)

This is fairly complicated, so take your time. Follow the stages on the diagram, which show the whole process for one follicle from start to finish.

The first stage of **oogenesis** takes place in the ovaries of a foetus before she is born. Cells on the outside of the ovaries divide by mitosis to make **oogonia**, which move towards the middle of the ovary and grow into primary oocytes. These are surrounded by a layer of follicle cells, which protect and nourish them, and are now called **primary follicles**. A girl has hundreds of thousands of these in her ovaries by the time she is born. Most deteriorate and only around 450 will mature fully during her reproductive life.

From puberty onwards, a primary follicle matures every month just before ovulation. It divides by meiosis, making a secondary oocyte and a polar body containing a spare set of chromosomes. The secondary oocyte begins its second meiotic cell division, but stops at metaphase. If it is fertilised by a sperm, the secondary oocyte completes meiosis to become a mature ovum, producing three polar bodies containing more spare sets of chromosomes.

Ovulation

A **Graafian (mature) follicle** is almost 1cm in diameter. It bulges out from the wall of the ovary just before ovulation. Fluid inside the follicle builds up pressure until it bursts, squirting the secondary oocyte out into the funnel-shaped opening at the end of the oviduct. This is the moment of ovulation, on around day 14 of the menstrual cycle. The ruptured primary follicle still has work to do; it turns into a **corpus luteum** (Latin for 'yellow body'), which makes the hormone **progesterone**.

Stages of oogenesis

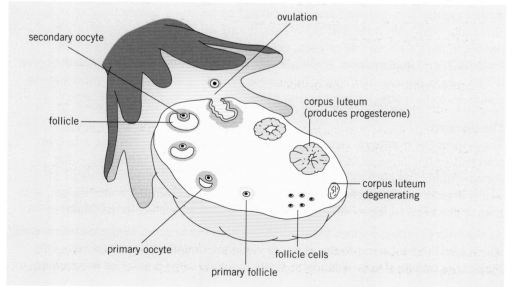

Follicle development in the ovary

Pregnancy and birth

Hormones

The interaction of hormones

Notice how the pituitary hormones stimulate the ovary to make oestrogen and progesterone and these regulate the pituitary hormones. This interaction is called feedback and regulates the menstrual cycle month after month.

- Follicle stimulating hormone (FSH) is released by the pituitary. It stimulates a primary follicle to start developing. A little luteinising hormone (LH) is also released, boosting the effect of FSH.

- The wall of the follicle starts making oestrogen, which causes the endometrium lining the uterus to thicken. It also temporarily inhibits the secretion of FSH.

- Oestrogen secretion peaks, causing a surge of LH with some FSH (which is no longer inhibited).

- The LH surge triggers ovulation.

- The corpus luteum makes progesterone, which stops the secretion of FSH and LH. It also maintains the endometrium.

- If fertilisation doesn't occur, the corpus luteum degenerates, ending the production of progesterone. The endometrium breaks down and then menstruation begins. Then, with no progesterone to inhibit them, FSH and LH are secreted again and the whole cycle starts over.

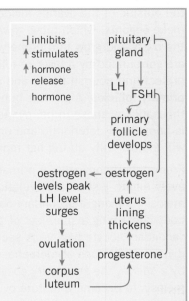

Pregnancy and birth

Foetal and maternal blood don't mix but the placenta allows solutes to pass between them, including:

- oxygen
- glucose
- amino acids
- fatty acids
- glycerol
- minerals
- vitamins
- hormones
- antibodies
- carbon dioxide
- urea

It also includes:
- nictoine
- alcohol
- heroin
- viruses.

Fertilisation

Human gametes must be protected against dehydration and this is achieved by sperm, swimming in semen, being released inside the vagina – **internal fertilisation**. The sperm travel through the cervix and uterus and along the oviduct. Of the millions that start the journey, only a few hundred sperm make it to the **secondary oocyte**. The acrosome on the front of the sperm releases enzymes that digest the ovum's membrane, allowing the head of the sperm to enter. Fertilisation occurs, producing a diploid **zygote**. This is moved along the oviduct by muscular contractions and cilia over about four days until it reaches the uterus, during which time the zygote mitotically divides to form a hollow ball of cells, called a **blastocyst**, which lodges into the endometrium – **implantation**. The blastocyst secretes **human chorionic gonadotrophin (HCG)**, which prevents the corpus luteum from breaking down, so that progesterone continues to be made, interrupting the menstrual cycle. Pregnancy has begun.

The placenta

The cells of the blastocyst keep dividing and some develop into the placenta that starts off as tiny projections called **chorionic villi** that embed themselves in the endometrium. The placenta takes over from the corpus luteum as after about 12 weeks it secretes **oestrogen and progesterone**, which maintain the endometrium, and **human placental lactogen**, which prepares the mammary glands for **lactation**.

As the embryo grows, so does the placenta. They are connected by the **umbilical cord**, which carries foetal blood to the placenta along the **umbilical artery** and back out though the **umbilical vein**. The foetus floats in a fluid-filled bag called the **amnion**.

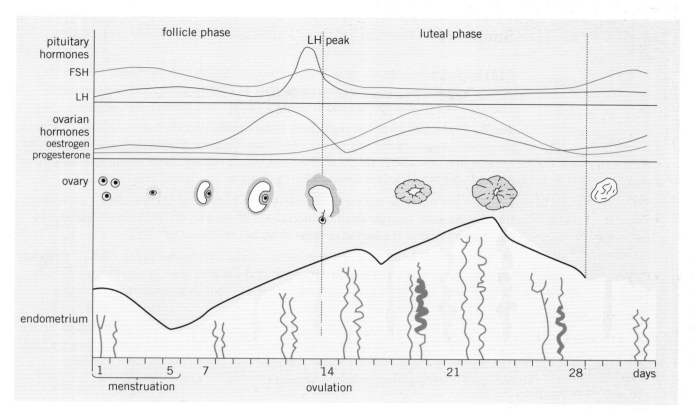

Hormones and their role in menstruation

Birth

Fourty weeks after implantation, the pituitary secretes **oxytocin** and together with prostoglandins, this causes the uterus to contract. The tension in the uterine muscle and pressure on the cervix, stimulates oxytocin secretion, causing the next contraction – **positive feedback**.

Lactation (milk production)

Oestrogen and progesterone prepare the mammary glands in the breasts for **lactation**, but production is inhibited by progesterone during pregnancy. Before birth, the progesterone levels fall and the pituitary secretes the **prolactin**, which triggers and sustains lactation. The stimulus of sucking on the nipple causes the pituitary to secrete oxytocin, which causes muscle contraction and milk is forced out of the breast. The milk made during the first few days of lactation is important: colostrum contains high concentrations of antibodies, which loan the baby some immunity against disease until its immune system is fully functional.

The placenta facilitates the exchange of substances between the mother and foetus

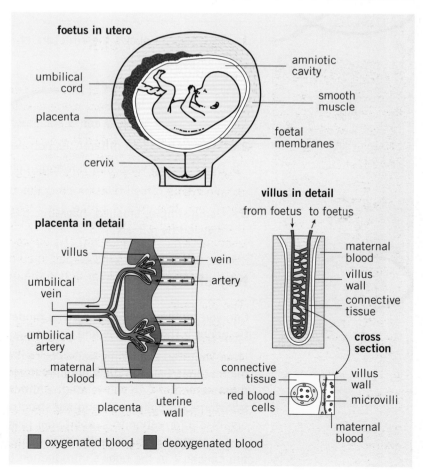

Summary

Cell division is needed for growth and reproduction. Mitosis produces two identical cells whose nuclei contain the same number of chromosomes as the parent cell. This type of cell division is used for growth, repair and asexual reproduction.

Asexual reproduction allows organisms to reproduce quickly but the individuals are all genetically identical. Many micro-organisms, plants, fungi and some animals use asexual and sexual reproduction. They reproduce rapidly under favourable conditions and are still able to generate variation to adapt to changing conditions.

Gametes are made by meiosis. This halves the number of chromosomes, preventing the chromosome number from doubling in each generation. Meiosis also produces genetic variation – the key advantage of sexual reproduction.

Male gametes tend to be small and are produced in large numbers. Female gametes tend to be large, containing food reserves and are produced in smaller numbers. Sexually reproducing organisms have organs to produce gametes and to enable fertilisation to occur.

Animals, such as humans, whose babies develop to an advanced stage before being born tend to have small numbers of offspring, which have a good chance of survival. The events of the menstrual cycle, pregnancy and birth are co-ordinated by feedback mechanisms involving hormones.

Practice questions

1   The diagram shows the structure of a human sperm.

  a Identify parts A–F.

  b Describe the functions of parts B and F.

  c A sperm is a haploid cell. Explain what this means.

2a State four typical differences between wind- and insect-pollinated flowers.

  b Describe two ways that flowering plants can avoid self-pollination.

  c What type of cell division produces ovules?

3   Using simple, labelled diagrams, describe the stages of one complete cell division by mitosis.

Key skills

Can you build a three-dimensional model to show the process of meiosis or mitosis? Go about it the right way and you can get credit for all parts of key skill PS3.

Have you tried examining pollen from the flowers of different plants under the microscope? Can you draw detailed diagrams and include the magnification and measurements? Can you relate the structure of the pollen to the way in which it is transferred in pollination and the structure of the receiving flower? Can you find and organise information to support your research? Do it the right way and you can demonstrate C3.1b, N3 (all), WO3 (all), LP3 (all) and PS3 (all) skills. Look through the checklist in the book's introduction to see what you need to do.

Exchange and transport

In this section, you will learn:

☞ how breathing works

☞ about blood and how the circulatory system works

☞ how the digestive system works

☞ how plants transport water and dissolved substances.

Life is very simple if you're a single-celled animal. Oxygen diffuses into you, carbon dioxide diffuses out and if you're feeling peckish you engulf a food particle and use a few enzymes to digest it. The point is, you have a large surface in relation to your volume. To put it another way, no part of an amoeba is far away from its surface. Even a multicellular organism, such as a flatworm can get away with using simple diffusion for gas exchange so long as it doesn't get too big.

Have you ever wondered why you don't find cockroaches the size of cats, like the ones you see on science fiction films? It's all to do with gas exchange. Insects are incredibly successful but they are restricted in size by not having lungs. They rely on gas exchange by diffusion, in and out of their tissues, using tiny tubes that open out to the air through holes called spiracles. This restriction is just as well for us, as there are far more species of insects than of all other animals put together.

There are clear benefits to being big:
- cells can differentiate to make specialist tissues and organs
- you can control your temperature better
- you can avoid being eaten by intimidating your predators.

Size causes problems. The bigger you are, the further away the cells in the middle of your body are from the outside. All those cells still need a steady supply of food and oxygen for respiration and they have to get rid of carbon dioxide and other waste. That's why big organisms need special gas exchange surfaces and mass transport systems to move substances around. Also, the more active an organism is, the greater the demand its cells have for oxygen and food and the faster waste will be produced. Humans have highly complicated and tightly integrated organ systems. Your digestive, breathing, circulatory and excretory system are all used in transport and exchange. As you will see, plants share similar problems and have evolved mass transport systems and gas exchange surfaces in response.

This is the biggest section in the book so take it slowly. Keep an eye on the examination board indicators, as not all the topics are required by every specification.

Breathing

KEY SKILLS N3

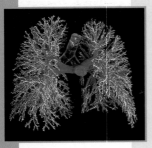

Resin cast of human lungs

Large, active organisms need a special respiratory surface for gas exchange. Gases could diffuse through our skin but the surface area is far too small to let enough oxygen through and dispose of the waste carbon dioxide quickly enough.

Lungs provide a large surface area for gas exchange as they contain thousands of tiny air sacs, called **alveoli**. Inhaled air enters the alveoli and oxygen diffuses across the thin walls into the blood. Carbon dioxide and water diffuse the other way, ready to be exhaled. In a sense, lungs are excretory organs, as part of their job is to get rid of the waste carbon dioxide and water produced by respiration.

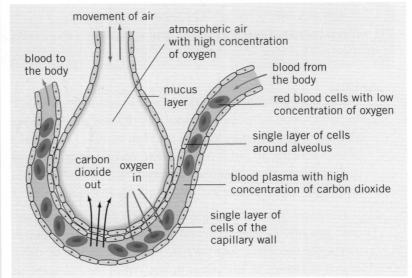

Gas exchange in the lungs

The alveoli are excellent at gas exchange as they are adapted to aid diffusion:

- the walls are thin and made of flat cells so gas molecules don't have far to diffuse
- oxygen dissolves in a moisture layer and diffuses through the walls, into the blood
- the moisture layer contains a **surficant**, that stops the walls of the alveolus sticking together by reducing the liquid's surface tension
- the network of capillaries brings CO_2 to the alveoli and takes O_2 around the body.

Breathing in and out gets rid of carbon dioxide, replacing it with oxygen, before it builds up in the alveoli. Together, these features work to maintain a steep concentration gradient. This is the difference in concentration of gases between the air in the alveoli and the blood that keeps the gases diffusing quickly. Even so, lungs need a huge surface area for gas exchange to happen quickly enough. Believe it or not, they have a combined surface area of about the size of a tennis court!

Fick's law shows the relationship between the factors that affect diffusion:

$$\text{Rate of diffusion} = \frac{\text{surface area x concentration difference}}{\text{thickness of membrane}}$$

Ventilation

When the pressure in the thorax is high, air is pushed out of the lungs. When it is low, the outside air pressure pushes air in. Two **pleural membranes** separate the lungs from the ribcage. The inner membrane is wrapped round the lungs and the outer membrane lines the inside of the ribcage and diaphragm. In between, a layer of **pleural fluid** acts as a lubricant, allowing the membranes to slide along each other as the lungs inflate and deflate without damaging the delicate lung tissue.

AS Guru™ Biology

Inspiration (breathing in)

- Diaphragm muscles contract and pull the diaphragm down, increasing the volume of the thorax and reducing the pressure.
- Deep breathing – external intercostal muscles pull the ribs up and out, increasing the thorax volume and reducing the pressure, so more air can enter the lungs.

Expiration (breathing out)

- The diaphragm relaxes and is pushed upwards. This reduces the thorax volume and increases the pressure. Air is forced out of the lungs.
- External intercostal muscles relax and the ribs move down and in. During exercise, internal intercostal muscles contract and pull the ribs further down and in. The pressure in the thorax increases, pushing air out of the lungs.

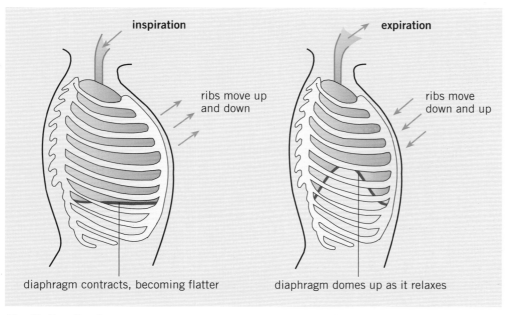

Ventilating the lungs

There are a few terms that you need to know, when learning about ventilation:

- **lung capacity** – the maximum amount of air that your lungs can contain
- **vital capacity** – the maximum useable volume of air
- **residual volume** – a small amount of air that remains in the lungs and airways
- **tidal volume** – the volume of air in a normal, resting breath
- **inspiratory reserve volume (IRV)** and **expiratory reserve volume (ERV)** are extra volumes of air you can breathe in and out during exercise, for example
- ventilation rate – the volume of air taken into the lungs in one minute. It's easy to work out: tidal volume x number of breaths taken per minute.

Control of breathing

Breathing is controlled by negative feedback, using nerve impulses instead of hormones, in the **medulla oblongata** (a region of your brain).

- Inspiratory cells in the medulla send nerve impulses to the diaphragm and external intercostal muscles. The muscles contract and you breathe in.
- The bronchi and bronchioles widen, which stimulates stretch receptor cells to send impulses to expiratory cells in the medulla. These send inhibitory impulses to the inspiratory cells, the diaphragm and intercostal muscles relax and you breathe out.
- The stretch receptors stop being stimulated and the expiratory cells stop inhibiting the inspiratory cells, so the diaphragm and intercostal muscles start contracting.

Carbon monoxide is an invisible, odourless gas, in cigarette smoke and car exhaust. It diffuses into the blood taking the place of oxygen on haemoglobin. Carbon monoxide poisoning, from poorly installed gas fires kills around 30 people per year in Britain.

Exhaled air contains lots of CO_2 and H_2O but little O_2. However, the level of N_2, which makes up nearly 80% of the air, stays the same.

GURU TIP

Try and think of breathing control in this way:

either, increased levels of carbon dioxide or decreased levels of oxygen, in the blood cause faster, deeper breathing.

Breathing

The ventilation cycle

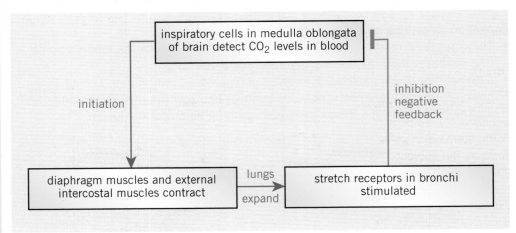

inspiratory cells in medulla oblongata of brain detect CO_2 levels in blood

initiation

inhibition
negative
feedback

diaphragm muscles and external intercostal muscles contract

lungs
expand

stretch receptors in bronchi stimulated

KEY SKILLS
C3, N3, IT3, WO3, LP3, PS3

The effect of exercise

Increased respiration in the muscles alters the chemistry of the blood, increasing the concentration of carbon dioxide and hydrogen ions. Cells called **chemoreceptors** in the medulla detect these changes and stimulate the inspiratory cells (and another group of cells called the ventral group). These, in turn, send extra impulses to the intercostal muscles and diaphragm, making you breathe faster and deeper.

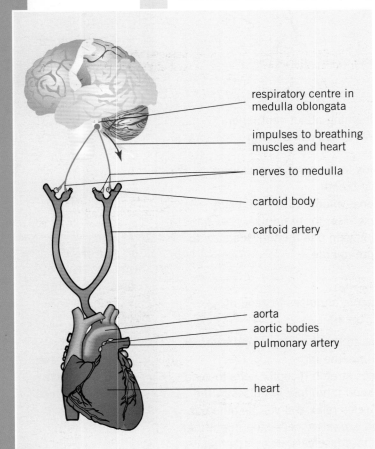

respiratory centre in medulla oblongata

impulses to breathing muscles and heart

nerves to medulla

cartoid body

cartoid artery

aorta
aortic bodies
pulmonary artery

heart

More chemoreceptors (situated in the **aortic bodies**) check the blood flowing through the aorta and the carotid arteries carrying blood to the head (the **carotid bodies**). If the concentrations of carbon dioxide and hydrogen ions are high, they signal the medulla to send more frequent impulses to the intercostal muscles and diaphragm. This causes faster, deeper breathing, increasing the supply of oxygen and getting rid of the carbon dioxide.

This is another example of negative feedback. Any increase in carbon dioxide level in the blood triggers a response (faster breathing), which reverses the change.

During exercise, if muscles can't get enough oxygen, they switch to anaerobic respiration. This releases less energy than aerobic respiration and produces lactic acid, which interferes with muscle contraction, giving you stitch. When you stop exercising, heavy breathing supplies oxygen that converts lactic acid to glycogen, paying off the 'oxygen deficit'.

Chemoreceptors detect changes in the blood

Measuring breathing

A **spirometer** can be used to measure ventilation by breathing through a tube into a sealed container with a hinged lid, which moves up and down as you breathe. The movement is recorded on a turning drum, called a **kymograph**. You can also use a position sensor plugged into a computer. Either way, you're supposed to get a graph showing changes in the volume of air in your lungs.

If you fit a cylinder of soda lime to the spirometer, it soaks up the exhaled carbon dioxide. The trace will show the amount of air in the spirometer going down as the subject steadily uses up the oxygen, and the carbon dioxide breathed out is removed by the soda lime. Working out oxygen consumption simply involves measuring how much the air volume falls in one minute.

Kymograph trace

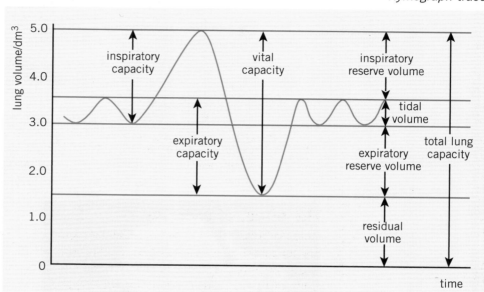

Spirometer experiment to measure ventilation

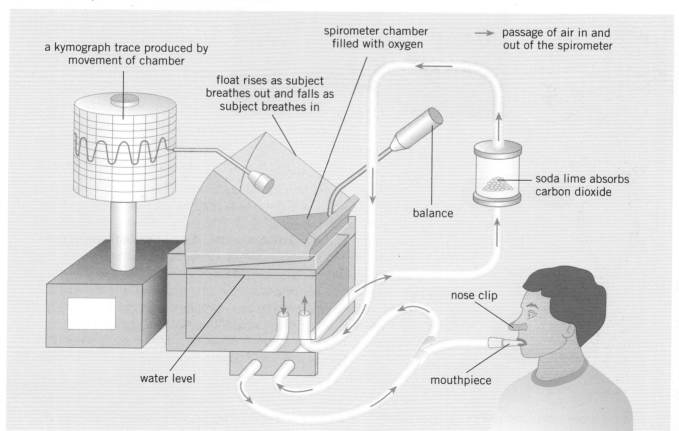

Circulatory system: blood

Blood carries out several vital jobs:

- transporting materials, such as food, oxygen, hormones, and waste products like carbon dioxide and urea
- distributing heat around the body to help maintain a constant temperature
- acting as a buffer – maintaining a constant pH by mopping up any excess acid or alkali
- providing pressure to allow some organs to work, for example filtration in the kidneys, formation of tissue fluid and erection of the penis
- defence against disease.

There are several components to blood, each adapted to carry out its particular function. You need to be able to recognise them on a microphotograph, name them and explain how they are suited to their various jobs.

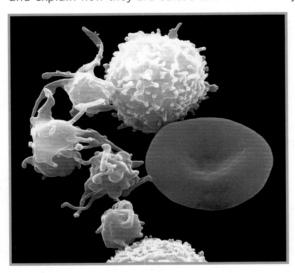

Coloured scanning electron micrograph of human blood showing a red cell, two white cells and four platelets

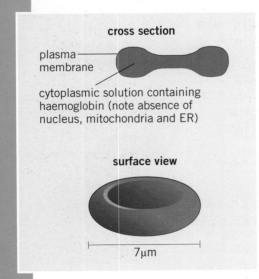

cross section

plasma membrane

cytoplasmic solution containing haemoglobin (note absence of nucleus, mitochondria and ER)

surface view

7μm

Red cells are biconcave discs

Red cells

The main job of red cells is to transport oxygen from lungs to respiring tissues. Also called **erythrocytes** (which just means 'red-cells'), they have a distinctive **biconcave disc** shape giving a big surface area in relation to their volume, so that oxygen can diffuse in and out quickly. They are packed full of **haemoglobin**, a red coloured globular protein, which latches on to oxygen in the lungs and releases it to cells as it travels round the body. To make room for as much haemoglobin as possible, they lack a nucleus, mitochondria and endoplasmic reticulum. Red cells are very small, as they need to squeeze through the narrowest capillaries, measuring just 7μm across (typical cells are normally around 40μm). They only last about 120 days, so new ones are constantly being made in the bone marrow.

If you live at high altitude for a long time, the body compensates for the relative lack of oxygen by making extra red blood cells.

White cells

White cells are concerned with fighting disease. They're sometimes called **leucocytes** which means 'white cells'. There are several sorts, but all have three features which make them easy to tell apart from red cells:

- they contain a nucleus
- they can be spherical or irregular in shape but are never biconcave discs
- they are nearly always much bigger than red cells.

White cells are divided into three main groups:

- Phagocytes (cell eaters) engulf and digest invading micro-organisms. Neutrophils are the most common, and can be spotted by their lobed nucleus and granular appearance.
- Monocytes have big, kidney shaped nuclei and develop into macrophages, the body's general purpose 'rubbish-clearing' cells.
- Lymphocytes secrete antibodies – chemicals which latch on to invading cells and destroy them. Lymphocytes have a big, round nucleus taking up most of the cell.

There are other white cells, called **eosinophils**, which protect the body against attack by larger, multicellular organisms, such as parasites. They secrete powerful enzymes, which break down the invader's cell walls. Also, less helpfully, they're also involved in allergic reactions.

GURU TIP
Make sure you've had a chance to identify these different blood cells under the micorscope or in micrographs. Questions asking you to tell them apart and say how they are adapted to their function, are common and an it's easy way to pick up marks.

Plasma

This is the pale yellow liquid part of blood, which carries dissolved substances. Exactly what it carries varies from one part of the body to another but could include proteins, salts, products of digestion, hormones, oxygen, carbon dioxide and urea. **Fibrinogen** is involved in the clotting mechanism, along with cell fragments called platelets. Plasma also carries heat around the body.

Tissue fluid and lymph

Some plasma seeps out through the walls of capillaries as they enter tissues and organs. This leaked plasma is called **tissue fluid** and it fills the spaces between your cells. It is very similar to plasma, without most of the plasma protein molecules, which are too big to get through. Tissue fluid bathes every cell, providing a liquid medium through which materials can pass between the blood and the cells. Some white cells can squeeze through tiny gaps in the capillary wall into the tissue fluid, where they clear up rubbish and hunt down any harmful microbes. Tissue fluid is constantly being re-circulated by seeping back into capillaries, which lead away from the organs. A small amount seeps instead into **lymphatics** – tiny, blind-ending vessels found throughout the body. The fluid, now referred to as **lymph**, slowly flows along the lymphatics, which join up to form larger lymph vessels. It eventually returns to the blood, into the big veins underneath the collarbone, called the **subclavian veins**.

Does your doctor ever grab you by the neck? Don't worry – it's just a way of seeing if your lymph nodes are swollen. If they are, it's a sign that your immune system is fighting an infection.

Lymphocytes, made in the bone marrow, build up in the lymph nodes, producing antibodies against the infecting pathogen.

The cardiovascular system

The mammalian blood system, or cardiovascular system, is made up of the heart and the blood vessels. Because blood stays inside the blood vessels as it circulates (unless you cut yourself), it's called a **closed blood system**.

Circulatory system: blood vessels

GURU TIP

'Pulmonary' is the word associated with lungs, 'hepatic' is to do with the liver and 'renal' is to do with the kidneys. Remember these and you've learnt the names of the major blood vessels.

The movement of blood around the body is an example of a **mass transport system**. Large amounts of materials are moved swiftly through a series of tubes. Anything bigger than an earthworm needs a mass transport system because diffusion only works quickly over very short distances.

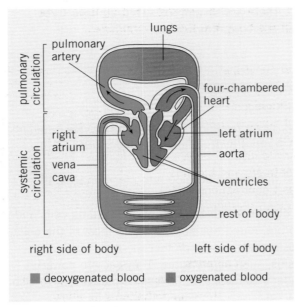

Mammalian circulatory system

If you follow blood all the way round, starting from the left ventricle, you'll find that it actually goes through the heart not once, but twice, before you get back to where you started. Mammals have a **double circulation**, with blood going round the body (the systemic circulation), through the heart, round the lungs (the pulmonary circulation), then back to the heart. This is probably because the amount of pressure needed to push blood all the way round the body would be too much for the lungs to handle.

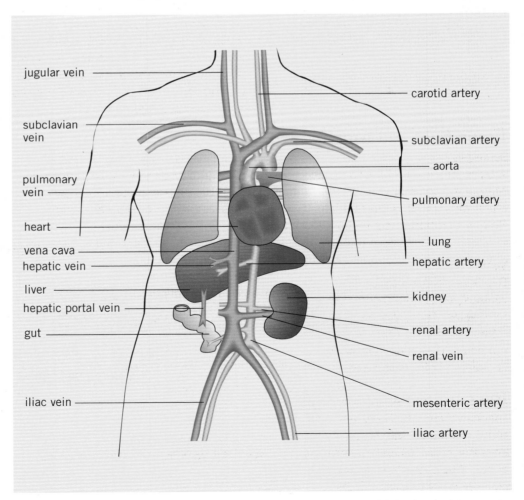

Main blood vessels in the human body

Blood vessels

Vessels carrying blood away from the heart are called **arteries**. Those taking blood towards the heart are **veins**. Between the two, and carrying blood right past the doorstep of almost every single cell in the body, are the tiny **capillaries**.

Arteries

Arteries carry blood quickly to the tissues at high pressure. They have a thick outer covering of tough collagen fibres (the **tunica externa** – literally 'outer coat'). The **tunica media** is a middle layer of smooth muscle and elastic fibres. The inner lining is made of thin, flat cells called **squamous epithelium**. The muscle layer of large arteries stretches when a ventricle contracts and blood is pushed out of the heart. Between heart contractions, the wall contracts, maintaining an even blood pressure. You can feel these surges as a pulse.

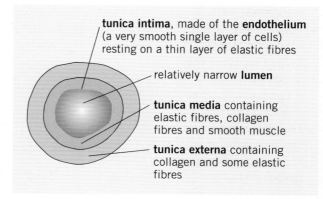

tunica intima, made of the **endothelium** (a very smooth single layer of cells) resting on a thin layer of elastic fibres

relatively narrow **lumen**

tunica media containing elastic fibres, collagen fibres and smooth muscle

tunica externa containing collagen and some elastic fibres

Transverse section of a small artery

Capillaries

Capillaries have walls that are just one cell thick (**endothelium**), to allow exchange of substances with the tissues. Their small diameter resists the flow of blood, slowing it down, allowing exchange to take place. Although the blood flows slowly through this network, there are so many capillaries that the total blood flow is large. They take blood as close as possible to all cells.

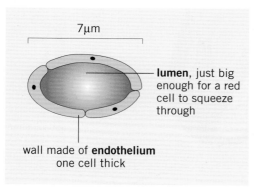

7µm

lumen, just big enough for a red cell to squeeze through

wall made of **endothelium** one cell thick

Transverse section of a capillary

Arterioles

Arteries branch into arterioles, which then branch into capillaries. Arterioles have muscular walls that can contract to restrict the flow of blood or relax to allow extra blood through. They play an important role in regulating the amount of blood flowing to different organs at different times.

Veins

Capillaries link up to form **venules**, which join together to form **veins**. These vessels return blood at low pressure to the heart. They have thin walls containing both elastic and muscle tissue, an inner lining of squamous epithelium and special valves, which prevent blood flowing backwards away from the heart. When you move about, the contracting muscles squeeze on the veins, helping to push the blood along.

Transverse section of a small vein

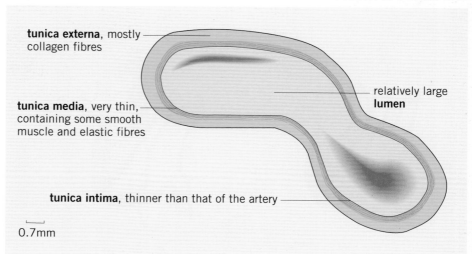

tunica externa, mostly collagen fibres

tunica media, very thin, containing some smooth muscle and elastic fibres

tunica intima, thinner than that of the artery

relatively large **lumen**

0.7mm

Exchange and transport

Circulatory system: the heart

The heart

Humans have a double circulation, so the heart is not one pump, but two. The right side pumps blood along the pulmonary artery around the lungs, the left side pumps blood through the aorta and around the body. The left side has the tougher job, so it is bigger and more muscular and the whole heart is about the size of a fist.

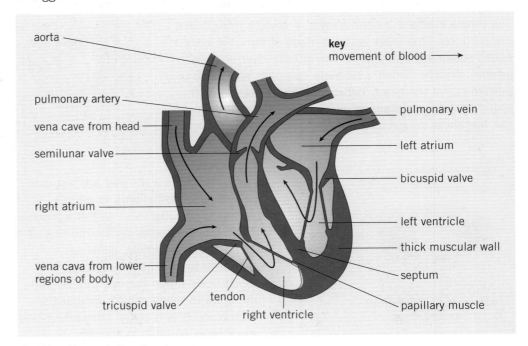

Section through the heart

Each pump has two chambers, an **atrium** at the top and a muscular **ventricle** at the bottom. **Atrioventricular valves** between the atria and ventricles, and **semi-lunar valves** where the arteries leave the heart, prevent blood flowing backwards through the heart. Remember, the pulmonary artery contains deoxygenated blood with a high concentration of carbon dioxide, and the aorta contains oxygenated blood with a low concentration of carbon dioxide.

The cardiac cycle

This is the sequence of events that takes place during one heart beat:

The cardiac cycle

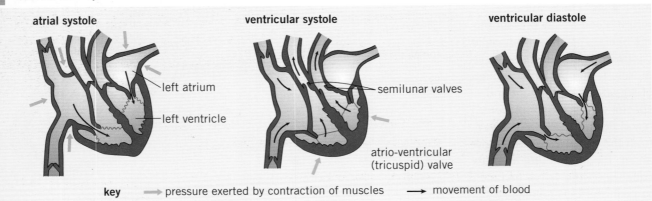

AS Guru™ Biology

- The atria fill with blood.

- The muscle in the walls of the atria contracts, pushing blood through the atrioventricular valves into the ventricles. This is called **atrial systole**. The semi-lunar valves stop blood going back out through the pulmonary veins and venae cavae.

- 0.1 seconds later, the powerful muscles of the ventrical walls contract, pushing blood out of the heart. This is **ventricular systole**, and lasts about 0.3 seconds. The atrioventricular valves shut, stopping blood going back up into the atria. If you look at the shape of the valve cusps, it's easy to see why this happens – the pressure of the blood against the cusps pushes them shut.

- The atria relax, letting the next lot of blood in from the veins.

- The ventricle wall muscles relax. The pressure of the blood in the arteries leaving the heart snaps the semi-lunar valves shut, preventing blood going back into the heart. This stage is **ventricular diastole**.

At rest, your heart completes this cycle about 70 times per minute.

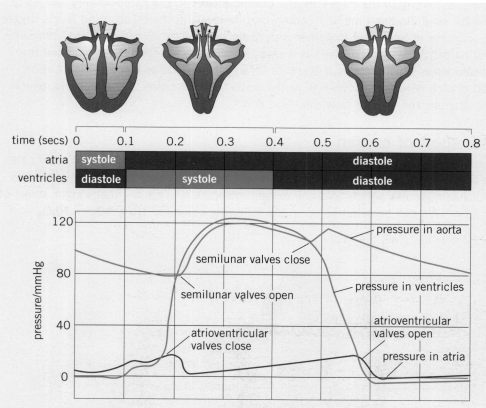

The cardiac cycle:

- contraction as pacemaker cells initiate systole

- the muscle walls squeeze, reducing the volume and increasing the pressure in the chambers, forcing blood in one direction

- the direction of blood flow causes the valves to open or close

GURU TIP

Study this carefully! Unlabelled versions are very popular with examiners. Try and remember these points:

- **systole** (contraction) is started by **pacemaker** cells

- chamber walls contracting reduces their volume, **increasing the pressure** inside

- the pressure of the blood opens and closes the valves so that blood flows through the heart in **one direction** only.

Exchange and transport

Pressure changes during the cardiac cycle

Controlling the heartbeat

Cardiac muscle is **myogenic** – this means it contracts and relaxes on its own, without needing signals from nerves. This rhythmical contraction needs to be co-ordinated for the heart to pump blood. A patch of muscle in the wall of the right atrium starts a heartbeat. It is called the **sino-atrial node** (**SAN** for short) and acts as the heart's pacemaker. The SAN sends a wave of electrical excitation across the walls of the atria, making them contract almost simultaneously. A band of non-conducting fibres at the bottom of the atria stops the wave from continuing down over the ventricles. Instead, the excitation wave is channelled along a patch of conducting fibres called the **atrioventricular node** (**AVN**). After a 0.1 second delay, the AVN passes the excitation down the septum, through the middle of the heart, along conducting fibres called the **bundle of His**, which quickly transmits the excitation wave to the bottom of the heart and out across the ventricles along the **Purkinje fibres**. As the wave spreads out, it makes the ventricles contract from the bottom upwards, pushing blood out through the arteries.

The SAN itself is controlled by a pair of nerves that bring impulses from the medulla. An accelerator nerve speeds up the heart and a decelerator nerve slows it down, according to the body's needs. **Chemosensors** in the carotid artery, the aorta and the medulla itself monitor carbon dioxide levels in the blood. High levels trigger an increase in heart rate (together with faster breathing). To prevent blood pressure getting dangerously high, there is a small swelling in the carotid artery called the **carotid sinus** which acts as a pressure sensor. It expands as blood pressure rises, and stretch receptors in the wall of the sinus send impulses to the medulla, which then signals the SAN to slow the heart down.

Transport of oxygen

Haemoglobin is brilliant at picking up oxygen in the lungs, carrying it to every cell in the body and releasing it to use for respiration. Each haemoglobin molecule has four haem groups, which hold one oxygen molecule (O_2) each – that's eight atoms of oxygen. This can be summed up in the following equation: **$Hb + 4O_2 \leftrightarrow HbO_8$**

> **GURU TIP**
>
> When it's combined with oxygen, haemoglobin is called **oxyhaemoglobin**. By the way, this formula drives chemists crazy because Hb isn't a proper chemical symbol (it's just shorthand for the very complicated haemoglobin molecule).

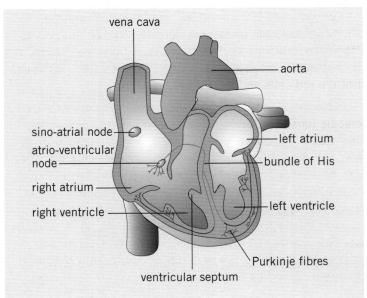

The SAN controls the heartbeat

The willingness of haemoglobin to combine with oxygen (its 'affinity' for oxygen) changes according to where it is. This is the clever bit – it has a very high affinity for oxygen in the lungs where it collects it, and a low affinity for oxygen in the respiring tissues where it needs to release the oxygen. The readiness of haemoglobin to let go of oxygen (to 'dissociate') in the presence of different concentrations of oxygen is shown by drawing a **dissociation curve**:

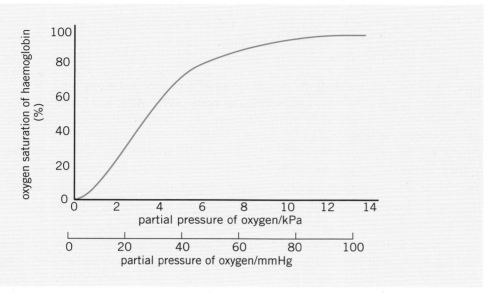

Haemoglobin dissociation curve

The **saturation of haemoglobin** tells you how loaded up the haemoglobin is. At 100% saturation, every single haemoglobin molecule would be carrying its full cargo of four oxygen molecules.

Partial pressure

Partial pressure is another way of describing the oxygen concentration around haemoglobin. **Kilopascals (kPa)** are a measurement of pressure – the higher the number, the more oxygen is available for the haemoglobin to bind to.

Notice how the percentage saturation of haemoglobin is low when the partial pressure of oxygen is low, but when there's plenty of oxygen, the percentage saturation is high. If you think about it for a moment, this is exactly what's wanted. In the lungs, the partial pressure of oxygen is high (about 12kPa) and the haemoglobin loads up with oxygen until it's around 97% saturated. When the haemoglobin gets to respiring muscle, the partial pressure of oxygen plummets to maybe just 2kPa (because the muscle's using it all up). Under these conditions, the haemoglobin lets go of three quarters of its oxygen so that it is only 25% saturated. This oxygen diffuses out of the red cells and into the muscle.

How does haemoglobin achieve the S-shaped dissociation curve? Think about an 'empty' haemoglobin molecule (one that isn't combined with any oxygen). Along comes an oxygen molecule, which combines with one of the four haem groups on the haemoglobin molecule. This distorts the shape of the whole molecule in such a way that it makes it easier for a second molecule to combine with a second haem group. When a second oxygen molecule combines, the same thing happens again, making it even easier for a third oxygen molecule to combine. This third one, however, makes it a little harder for a fourth (and last) oxygen molecule to combine with the final haem group. The upshot is a dissociation curve that is very steep in the middle. A small change in the partial pressure of oxygen causes a very big change in the amount of oxygen carried by the haemoglobin.

GURU TIP
Haemoglobin dissociation curves are popular in exam questions. Make sure you understand what the axes mean, why the curve is S-shaped and why this shape helps haemoglobin to do its job so well.

Exchange and transport

The Bohr shift

Haemoglobin releases oxygen even more readily where there is a high partial pressure of carbon dioxide. This is the situation in respiring tissues, such as muscle, which are churning out carbon dioxide as a waste product of respiration.

Red blood cells contain an enzyme called **carbonic anhydrase**. This turns carbon dioxide and water into carbonic acid (H_2CO_3), which splits up straightaway into hydrogen ions (H^+) and hydrogen carbonate ions (HCO_3^-). Haemoglobin combines easily with these hydrogen ions, forming **haemoglobinic acid** (HHb) and giving up the oxygen it is carrying in the process. This achieves two things:

- the haemoglobin acts as a **buffer**, keeping the blood at an almost neutral pH by mopping up hydrogen ions which would otherwise turn the blood acid
- haemoglobin gives up its oxygen more easily wherever there is a lot of carbon dioxide about. This is called the **Bohr effect**, after the person who discovered it.

The dissociation curve for haemoglobin at high partial pressure of carbon dioxide is to the right and below the curve at low partial pressure. Haemoglobin lets go of oxygen more easily in the presence of carbon dioxide – just what is needed.

In the lungs, these reactions go in reverse, releasing carbon dioxide, which diffuses into the alveoli and leaving the haemoglobin free to collect fresh supplies of oxygen.

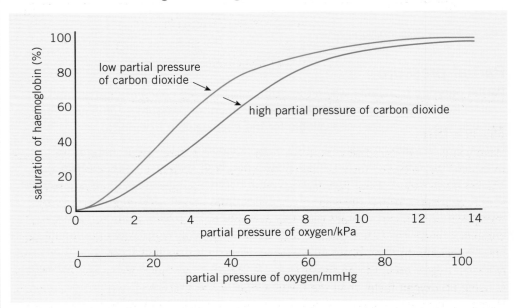

Dissociation curves showing the Bohr effect

Carbon dioxide transport

- About 85% of the carbon dioxide made by respiring cells is carried in the form of hydrogen carbonate ions, which are made in the red cells (see Bohr effect above) and then diffuse out into the plasma.
- Around 10% combines directly with haemoglobin, forming carbamino-haemoglobin.
- The other 5% just dissolves in the blood plasma.

GURU TIP

Can't remember which way the Bohr shift moves the dissocaition curve? Try this: 'Bohr' has the letters b and r in it. This effect shifts the curve below and to the right.

KEY SKILLS N3

Foetal haemoglobin

A developing foetus has problems receiving oxygen from the mother's blood. The foetal blood in the placenta has a low partial pressure of oxygen (because it is respiring), but only a little lower than the mother's blood. If the foetus' haemoglobin had the same dissociation curve as the mother's, it wouldn't be able to collect oxygen quickly enough. Instead, foetal haemoglobin has a higher **oxygen affinity** than normal haemoglobin, in other words it readily picks up oxygen that adult haemoglobin has dropped. For any given partial pressure of oxygen, foetal haemoglobin is slightly more saturated than adult haemoglobin. Its dissociation curve is above the curve for normal haemoglobin. After birth, the baby replaces its foetal haemoglobin with normal haemoglobin.

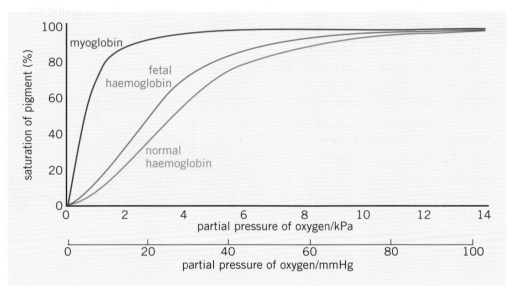

Dissociation curves for haemoglobin, myoglobin and foetal haemoglobin

GURU TIP
In a muddle with these curves? The word 'foetal' has an a and an l in it. The curve for foetal haemoglobin is above and to the left of the adult haemoglobin curve.

Myoglobin

Have you ever wondered why the meat on a chicken is dark in some places and white in others? White breast muscle makes the wings flap, which is a bit pointless as chickens can't fly, but the dark leg muscles work harder and need a guaranteed supply of oxygen, especially when trying to out-run a fox! The colour comes from myoglobin, a dark red pigment (like haemoglobin but with one haem group), which combines reversibly with oxygen. It acts as an emergency reserve of oxygen because, once myoglobin has combined with oxygen, it forms a very stable compound. It only releases oxygen when the partial pressure is very low, in other words when the muscle is using up oxygen faster than the haemoglobin in the blood can bring fresh supplies. Human muscle tissue also contains myoglobin, especially the arm and leg muscles which can create a particularly high oxygen demand during exercise.

High altitude

The partial pressure of oxygen at 6000m is half that at sea level. However quickly and deeply you breathe at this height, haemoglobin will never be more than about 70% saturated with oxygen. This can lead to altitude sickness, which can be lethal. The brain's arterioles widen to increase blood flow, causing fluid to leak from the capillaries into the brain tissue. This leads to dizziness, confusion and loss of consciousness. People (and animals) accustomed to living at high altitude have more red blood cells, with extra haemoglobin, bigger lung capacity and more powerful pulmonary circulation.

Climbers must spend several weeks at high altitude before going up the highest mountains to get acclimatised, giving their bodies time to produce extra red blood cells.

Exchange and transport

Digestive system

KEY SKILLS
C3, N3, IT3, WO3, LP3, PS3

An orange, probably covered in Penicillium

This technique can be used to measure the activity of other enzymes. You could mix white protein powder into the agar, instead of starch. Protease enzymes would digest the protein, turning the cloudy agar clear.

Extracellular digestion in fungi

Fungi, such as mushrooms and moulds, feed on the dead remains of plants and animals. This is called **saprophytic nutrition** (or saprobiotic, which means the same thing). The body of a mould is made up of very thin threads called hyphae that spread through the food source forming a network called a **mycelium**, providing a huge surface area for a small volume. The hyphae secrete enzymes, which diffuse through the cell wall, into the food. Here, they break down big, insoluble molecules into small, soluble ones that can be absorbed into the **hyphae** by facilitated diffusion and active transport. Because digestion happens outside of the mould's cells, it is called **extracellular digestion**.

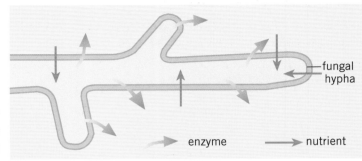

Extracellular digestion by a fungus

Measuring enzyme activity

Digestion depends on enzymes. If you wanted to make a biological washing powder that would be good at digesting starch-based stains, you might want to compare different fungi to see which produce the most starch-digesting carbohydrase enzymes. A simple technique is to use a starch agar plate. Starch is mixed into hot, liquid agar, which is then poured into a petri dish and left to set. Standard-sized samples of the different fungi are put onto the agar and left for a few hours. Iodine solution is poured onto the plate, staining the areas that contain starch blue-black. Areas around the fungi, where the starch has been digested show up as clear circles – the bigger the circle, the more effective the enzyme has been. If you include a standard sample containing a known amount of carbohydrase, comparing the circles allows you to work out how much enzyme the unknown samples contain. Such a technique, based on comparison with a standard sample, is called an **assay**.

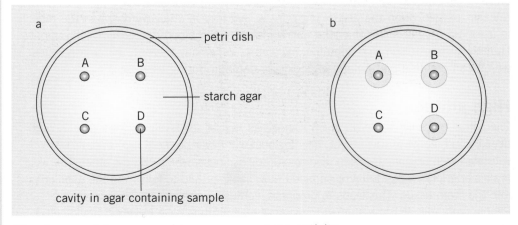

Starch agar plates are used to measure enzyme activity

Digestion in humans

Bigger organisms, such as humans, need a specialised digestive system. You'll remember from GCSE that food is a complex mixture of chemicals that the body needs for energy, growth and repair. Some chemicals, such as vitamins and minerals, can be absorbed as they are. Most food chemicals – carbohydrates, lipids and proteins – are in the form of big, insoluble molecules that need to be broken down by hydrolysis into small, soluble molecules before they can be absorbed. They need to be digested. As food moves through the digestive system, it undergoes a series of physical and chemical processes that achieve this goal.

The gut wall has three layers:

- an outer layer of muscle
- a middle layer called the **submucosa**
- an inner layer called the **mucosa**

Food travels along a space called the **lumen**. Each part of the alimentary canal is adapted to carry out its particular function in digestion.

> **GURU TIP**
> Chewing (and the churning action of the alimentary canal) is called mechanical digestion. The action of digestive enzymes breaking down food molecules is called chemical digestion.

> Starch is the main carbohydrate that we eat. Plants make lots of starch as an energy store, and we are well adapted to digesting it. Cellulose is another common plant carbohydrate but we don't make an enzyme to digest it. Instead, it provides the fibre that gives bulk to food and helps to push it along the gut.

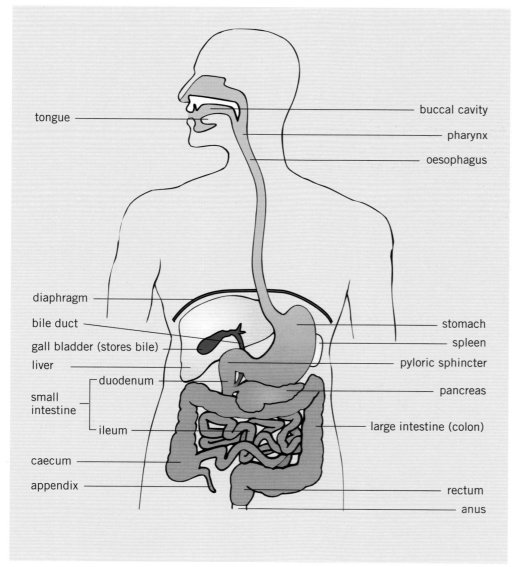

Human digestive system

Exchange and transport

Digestive system

The mouth
Mastication (chewing) breaks up food, increasing the surface area for enzymes to work on and making it easy to swallow. Saliva from the salivary glands lubricates the food and contains amylase, a carbohydrase, which starts to break down starch into maltose.

The oesophagus
This has two thick layers of outer muscle, which uses waves of contraction called **peristalsis** to push food along. Glands secrete slippery mucus and the mucosa is lined with layers of flat cells that can rub off without damage.

The stomach
Three layers of muscle, running in different directions, contract and relax to produce a churning action, which mashes up the food. The thick mucosa is full of deep pits called gastric glands that secrete mucus, hydrochloric acid and a protein-digesting enzyme called **pepsin**. The acid kills microbes and provides the low pH that pepsin needs to hydrolyse proteins. Pepsin is called an endopeptidase because it breaks bonds between amino acids inside long polypeptide chains ('endo' means 'inside'), chopping them up into shorter polypeptide chains.

The duodenum
This is the first section of the small intestine. Two layers of muscle move food along by peristalsis. **Brunner's glands**, in the submucosa, secrete alkaline mucus, which helps to neutralise stomach acid. The mucosa is folded up and covered with lots of little projections called villi to increase the surface area, ready to begin absorbing the products of digestion. Many tiny glands secrete mucus and enzymes. The pancreatic duct brings pancreatic juice from the pancreas. This contains three groups of enzymes:

- carbohydrases, for example amylase, which completes the breakdown of starch into maltose started in the mouth
- lipases, to break down fats and lipids into fatty acids and glycerol
- **endopeptidases**, to continue breaking down proteins into shorter polypeptides and **exopeptidases** to cut single amino acids off the ends of these polypeptides.

More exopeptidases on the surface of the microvilli complete the digestion of protein. The two protein-digesting enzymes work well together. The endopeptidases cut up long polypeptide chains into shorter lengths, leaving lots of ends for the exopeptidases to work on. The amino acids that are snipped off the ends of the polypeptides by the exopeptidases are then absorbed into the epithelial cells by facilitated diffusion.

The liver
This organ makes bile, which is stored in the gall bladder. When food arrives from the stomach, bile is released into the duodenum. It neutralises the stomach acid, turning the contents of the duodenum slightly alkali – conditions that suit the enzymes here. Bile also **emulsifies** fat, breaking it up into tiny droplets with a big surface area for lipase to work on.

GURU TIP
Imagine that you are a chicken sandwich. Describe your journey through the digestive system, describing what enzymes are breaking down your carbohydrate (bread), protein (chicken), and lipid (butter). This will help you to visualise and recall the stages in digestion.

To avoid digesting itself, the stomach secretes an inactive form of pepsin, called pepsinogen. This only gets converted into active pepsin in the acidic conditions of the stomach lumen. The mucus lining the stomach wall also provides a protective barrier.

The ileum

Similar to the duodenum, the second part of the small intestine is where digestion is completed and the products are absorbed. The epithelial cells lining the villi are covered in microvilli, increasing the surface area. The sub-mucosa has a rich supply of blood and lymph vessels to collect the products of digestion and carry them away. Maltase, in the outer membrane of the surface cells of the microvilli, finishes carbohydrate digestion by breaking down maltose into glucose, which is taken into the epithelial cells by active transport. Amino acids are also absorbed by active transport. Salts and vitamins pass into the epithelial cells by diffusion. Fatty acids and glycerol diffuse into the epithelial cells and on into lymph capillaries called lacteals in the middle of the villi.

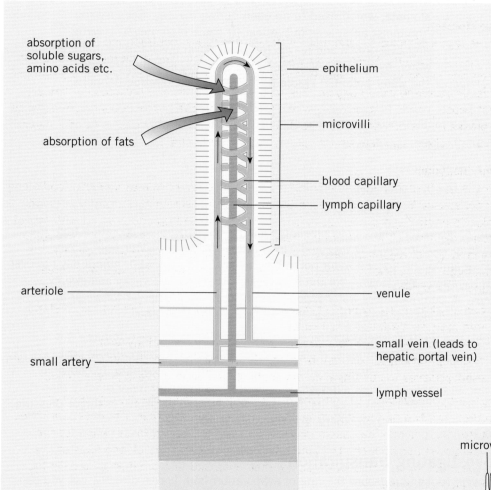

Transverse section of a villus and a close-up of a single epithelial cell

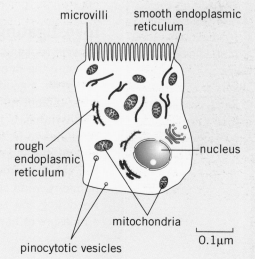

> **GURU TIP**
> Strictly speaking, defecation is not excretion. The undigested material in faeces has never been part of the body, it has simply passed through it. The correct term is **egestion**.

> The large intestine absorbs water from undigested material, which is then stored in the rectum ready to be expelled through the anus. It also plays host to over 400 species of bacteria! Some of these are vital, producing products, such as vitamin K and folic acid – the others rarely do any harm.

Exchange and transport

Plant transport systems

Plants have two transport systems. One carries water and inorganic ions from the roots to the parts of the plant above the ground, while the other carries organic substances, made by photosynthesis, from the leaves to the rest of the plant. Simple diffusion takes care of the transport of carbon dioxide and oxygen.

Transpiration

Plants lose water through their leaves. Photosynthesis uses carbon dioxide from the air to make oxygen and both these gases have to diffuse in and out of the leaves, through the **stomata**, which need to be open during the day. Because the air outside is drier than the moist air spaces in the leaf (it has a more negative water potential), water evaporates and leaves the plant down a water potential gradient. This loss of water, by evaporation through the stomata, is called transpiration. The steeper the **water potential gradient**, the faster transpiration happens.

Three factors increase the rate of transpiration:
- increased temperature
- lower humidity
- increased wind speed.

Warmth gives the water molecules more energy, encouraging evaporation from the liquid to vapour phase. Dry air and wind remove the water vapour from around the leaf, maintaining a steep water potential gradient. On a hot, still, humid day however, transpiration slows down – for the same reason that you feel sticky and sweaty. Warm air can hold more water vapour than cold air, so it's harder for water to evaporate unless there's a breeze to carry the saturated air away.

Light also tends to speed up transpiration. As the stomata open to allow gas exchange for photosynthesis, more water evaporates as an unavoidable cost. At night, photosynthesis stops and most plants shut their stomata, to conserve water and replenish their supplies. If a plant is losing water more quickly than it can replace it from the roots, its cells lose the outward pressure, or **turgor pressure**, that maintains their rigid shape and the plant wilts.

Investigating transpiration

It's very tricky to measure the rate of evaporation from leaves directly, but you can measure the rate of uptake of water by a shoot quite easily with a **potometer**. This gives a good idea of how fast transpiration is taking place, as most of the water a plant takes up is lost by transpiration, although a little will be used for photosynthesis. There must be no air bubbles trapped in the apparatus, so it's a good idea to fit the shoot into the potometer underwater. Cutting the stem underwater and at an angle seems to help prevent air bubbles from sticking to it. You can then expose the shoot to different sets of conditions (temperature, humidity, wind speed and light), to see how they affect transpiration rate. Recording how fast the water meniscus moves along the capillary tube gives you a measure of how fast transpiration is happening.

GURU TIP
An easy way to remember what conditions speed up transpiration is to think what sort of day you'd choose to hang out washing: dry, hot and breezy. These are exactly the sort of conditions that make plants lose water more quickly.

KEY SKILLS
C3, N3, IT3, WO3, LP3, PS3

Plants face a dilemma. Big, flat leaves with loads of stomata for gas exchange are great for photosynthesis, but bad news for water loss. See what kinds of strategies plants use to get around this paradox.

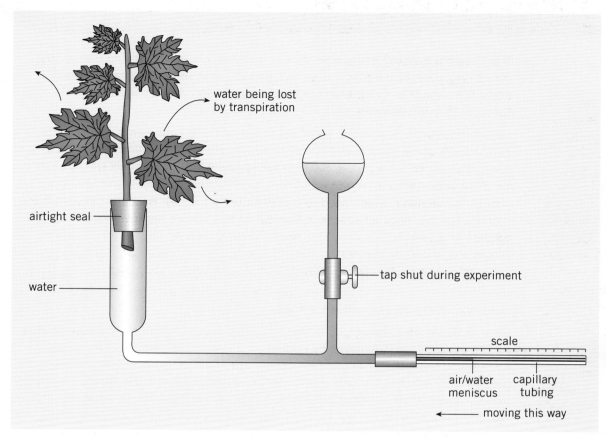

Potometer, measuring the rate of water evaporation

Plants normally get their water from the soil through long, thin extensions of the epidermis just behind the root tip, called **root hairs**. These provide a large surface area to aid the movement of water through the plasma membrane into the cytoplasm and vacuoles of the root cells. Although soil water contains some dissolved minerals, it is a very dilute solution with a high water potential. The cytoplasm, on the other hand, has a fairly high concentration of mineral ions and sugars, giving it a relatively low water potential. So water goes into the root cells down a water potential gradient.

Into the root

The water needs to get from the root hair cells across the **cortex** and into the xylem vessels, ready to move up the plant. Xylem has a lower water potential than the root hairs (you'll see why in a minute), so again water moves down a water potential gradient. There are two routes it can take.

The route through non-living tissue is called the **apoplast pathway**. The cortex cell walls are made of layers of criss-crossing cellulose fibres that act like blotting paper. Water is drawn along from the wall of one cell to the next, through the intercellular spaces, without ever actually entering the cells.

Symplast and apoplast pathways

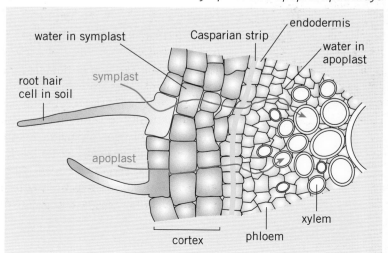

Plant transport systems

GURU TIP
Remember that xylem tissue has two functions:

- supporting the plant
- transporting water and mineral ions.

The **symplast pathway** is the route that water takes through the living parts of the cortex cells: the membranes, cytoplasm, and vacuoles. It moves from cell to cell along tiny strands of cytoplasm called **plasmodesmata**, which pass through the cell walls, linking neighbouring cells.

Before it can enter the xylem, water has to get past the **endodermis**. These cells have a thick, waxy band of waterproof **suberin** in their walls, which blocks the apoplast pathway. This barrier is called the **Casparian strip**. The only way to cross the strip is through special passage cells. It is thought that plants use this as a 'customs checkpoint', controlling which mineral ions are allowed into the xylem.

Mineral ions

It seems that mineral ions get into the root by active transport, often against a concentration gradient. They then pass into the xylem and are carried in solution to the leaves and growing parts, where they move to wherever they are needed by a mixture of diffusion and active transport.

Good evidence for mineral ions travelling up xylem comes from **ringing experiments**. All the living tissue is removed by cutting off a ring of bark around a woody stem, just leaving the xylem. The plant's roots are put in a solution containing phosphate ions labelled with radioactivity. After a while, a Geiger counter can be used to show that the labelled phosphate has travelled up to the leaves, through the xylem.

Lignin is very tough and gives plants their 'woodiness'. It is strong, flexible and waterproof. This makes it ideal for support and for making tubes to transport water.

Up the stem

Xylem tissue transports water and mineral ions and also provides support. Xylem is the stringy stuff that gets stuck in your teeth when you eat raw celery.

> **Remember:** a tissue is a group of cells that work together to carry out a job, such as transporting water and supporting the plant.

These are the types of cell in the xylem tissue of a flowering plant:

- **vessel elements** – water transport cells
- **fibres** – long, dead and lignified (woody) cells that provide support
- **xylem parenchyma cells** – normal plant cells, which act as packing tissue, containing the usual organelles apart from chloroplasts, as there isn't enough light for photosynthesis.

Vessel elements start off as long, thin cells arranged end to end. A tough, waterproof substance, called **lignin**, is laid down in the cell walls, apart from where adjacent cells join and the walls break down. The contents of the cells die, leaving a long, hollow tube.

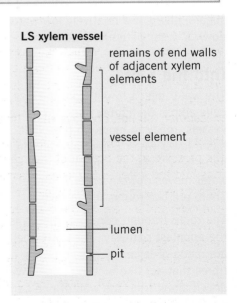

LS xylem vessel

remains of end walls of adjacent xylem elements

vessel element

lumen

pit

Xylem vessels transport water and mineral ions

Where there were once plasmodesmata in the walls, pits are formed. These are tiny gaps crossed by cellulose cell walls that water can pass through. This is how water gets into the xylem in the root.

Bundles of xylem vessels run along the middle of roots. In the stem, they form groups, together with phloem, called vascular bundles.

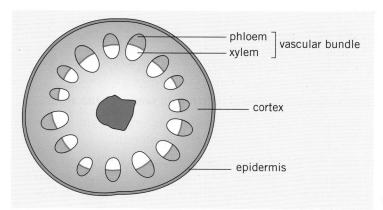

phloem
xylem ⎤ vascular bundle
cortex
epidermis

Transverse section of a stem

The cohesion-tension mechanism

Water is constantly evaporating from the leaf (transpiration). This lowers the pressure of the water in the xylem in the leaf. The higher pressure of the water in the root xylem pushes water up the stem, replacing the water being lost from the leaf. It's a bit like sucking on a straw: lowering the pressure at the top (by sucking) makes liquid move up the straw, pushed by the higher pressure acting on the liquid in the tumbler. The important thing to understand is that a column of water in a closed tube is almost impossible to break.

There are strong **cohesion forces** holding the water molecules together, and water doesn't stretch. If you pull on one end, the whole column moves. So long as air doesn't get into the tube, the water moves up in a continuous stream – an example of a **mass flow system**. Air locks interfere with this process, which is why you should cut the bottom off fresh flower stems before you put them in to water. Because the water is under tension, with gravity pulling down against the upward force caused by transpiration, this effect is called the **cohesion-tension mechanism**. There are also adhesion forces attracting water molecules to the lignin walls of the xylem vessels, helping the water to move as a continuous column. The tension on the water column tends to pull the walls inwards – the lignin is needed to stop the vessels collapsing. This effect is strong enough to make a tree trunk narrow measurably on hot, dry days when transpiration is happening quickly.

Plants can encourage the flow of water up the stem in an active way, by increasing the pressure at the bottom of the xylem vessels. They secrete solutes such as mineral ions into the water in the xylem vessels in the root (using active transport). This lowers the water potential inside the xylem, drawing in water from the neighbouring root cells and increasing the pressure, helping to push water up the xylem. This active process is called **root pressure**. It is probably less important than the cohesion-tension mechanism, as water still goes up a plant's xylem even when it is dead (cut flowers don't wilt immediately so long as the ends of the stems are put in to water).

To put it in a nutshell, water moves into the roots, up the stem and out the leaves down a water potential gradient. This is sometimes called the **transpiration stream**.

GURU TIP
The energy needed to pull water up the stem in this way comes from the sun. It's the water evaporating from the leaves that sets up the water potential gradient that makes water flow up the xylem.

Exchange and transport

Plant transport systems

Sieve elements are different from xylem vessel elements in three ways, helping them transport materials up and down the stem:

1 they don't have lignified walls, so water and solutes can pass in and out

2 they are living cells with cytoplasm, containing all the usual organelles, apart from nuclei

3 they have companion cells alongside them, stuffed full of mitochondria.

Xerophytes

These are plants that are adapted to survive in very dry conditions, such as deserts and coastal sand dunes. They have all sorts of cunning tricks to gather as much water as they can and then hang on to it, losing as little as possible by transpiration. The important features to remember are:

• roots that spread out over a big area just below the surface, ready to make the most of any rain

• swollen stems to store water, with a waxy epidermis to prevent evaporation

• leaves adapted to form spines to stop thirsty herbivores eating them, and stems with chloroplasts to take over the job of photosynthesis

• being covered in hairs to trap moist air and reduce evaporation

• stomata sunk into pits which trap still, moist air to slow transpiration

• leaves which curl up in hot, dry weather to slow down transpiration

• taking in CO_2 at night so that they can close up their stomata in the hot day to reduce water loss, but still photosynthesise using the stored CO_2.

Because of the inescapable trade-off between gas exchange for photosynthesis and water loss by transpiration, xerophytes, like cacti, tend to grow very slowly.

Translocation

Carbohydrates made by photosynthesis have to be transported around a plant, to where they are needed for growth, respiration or storage. This movement of dissolved organic substances from one place to another happens in the phloem and is called **translocation**. Unlike the movement of water and inorganic mineral ions in xylem, which is always in the same direction (upwards from roots to leaves), ringing experiments have shown that phloem sap can flow either way – even in the same vascular bundle. Think about a potato plant in the summer: the leaves are photosynthesising, making sugar, and this has to be translocated down to the potato tubers to be turned into starch and stored. When it's time for new growth, enzymes turn the starch back into sugar, which now has to be moved in the opposite direction, away from the potato. So a potato tuber can be either a **source**, loading sucrose into the phloem, or a **sink**, taking it out.

Phloem and its role in translocation

Phloem tissue is made of sieve elements, companion cells, parenchyma and fibres.

Sieve elements join end to end to make **sieve tubes**. Each element is a living cell, with only a small amount of cytoplasm and no nucleus or ribosomes. The **sieve plate** at the junction between two sieve elements has lots of holes in it, which can let liquid through. Companion cells lie alongside the sieve elements. They are fairly normal plant cells, but with extra mitochondria and ribosomes. They have lots of plasmodesmata linking their cytoplasm with the cytoplasm of the sieve elements. It is thought that these very metabolically active cells are involved in loading and unloading the sieve elements with sucrose.

The enlarged stem of a Cactaceae *cactus, stores large quantities of water, enabling it to survive in arid regions*

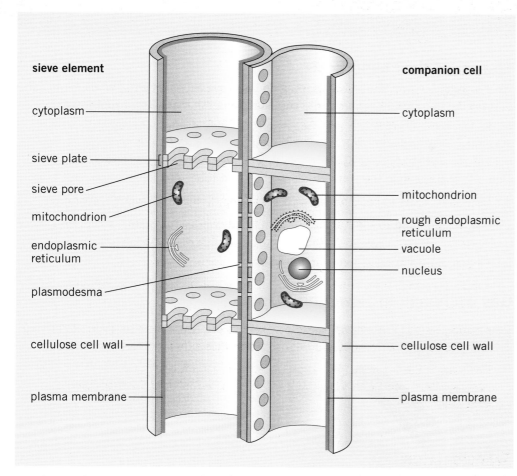

sieve element
- cytoplasm
- sieve plate
- sieve pore
- mitochondrion
- endoplasmic reticulum
- plasmodesma
- cellulose cell wall
- plasma membrane

companion cell
- cytoplasm
- mitochondrion
- rough endoplasmic reticulum
- vacuole
- nucleus
- cellulose cell wall
- plasma membrane

Sieve tube element and companion cell

What's in phloem sap?
Collecting a sample isn't easy; if you cut a stem, the exposed ends of the sieve elements quickly seal – a bit like blood clotting if you cut yourself.
One trick is to let an aphid stick its needle-like stylet into a sieve tube. The pressure of the sap pushes it out of the sieve tube and into the aphid; it doesn't even have to suck. Just when its enjoying a nice drink of sap, chop it off! Sap will carry on coming out of the end of the stylet, ready to be collected and analysed.

Loading and unloading the phloem

Some of the triose sugars made by photosynthesis in the leaf mesophyll cells are turned into sucrose, ready for translocation. Dissolved in water, this moves towards the phloem through either the symplast pathway (cell to cell via plasmodesmata) or the apoplast pathway (along cell walls). Then the sucrose is pumped into the companion cells by active transport and on into the sieve elements through the interconnecting plasmodesmata. At the sink end, sucrose probably leaves the phloem by diffusion. A concentration gradient is maintained by enzymes, which modify the sucrose. For example, **invertase** hydrolyses sucrose into fructose and glucose in a growing fruit. In a potato tuber, the sucrose is turned into starch for storage. In a growing leaf, it would be used to make cellulose for new cell walls. The point is that the sucrose is used and doesn't build up, keeping its concentration low in the sink organ compared to the phloem.

Sources, sinks and mass flow in phloem

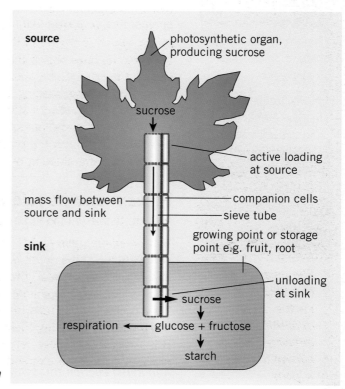

source
- photosynthetic organ, producing sucrose
- sucrose
- active loading at source
- mass flow between source and sink
- companion cells
- sieve tube
- growing point or storage point e.g. fruit, root
- unloading at sink

sink
- sucrose
- respiration ← glucose + fructose
- starch

The mass flow hypothesis

What actually moves the phloem sap along the sieve tubes? It's not as if plants can pump it up and down. The favourite explanation is the **mass flow hypothesis**.

At the source end (usually a leaf), sucrose is actively loaded into the sieve elements. This decreases the water potential in the sap, so water diffuses in from the surrounding cells. You can see how this will increase the pressure as the sieve element fills, pushing the solution along the sieve tube. At the sink end (say, a growing bud), sucrose is taken out of the sieve elements, taking water with it. The result is movement of the entire column of sap along the sieve tube from source to sink, carrying any dissolved solutes along with it. Another example of mass flow, but this time as an active process that involves using energy (for active transport), unlike passive water transport in dead xylem.

Gas exchange in flowering plants

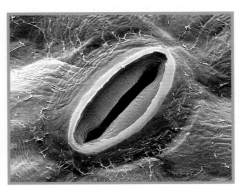

Leaves are adapted to be good at carrying out photosynthesis, preferably without losing too much water. This means they need a big surface for capturing light, plenty of photosynthetic cells stuffed with chloroplasts, stomata for gas exchange and a network of phloem and xylem vessels for transporting materials to and from the rest of the plant.

Scanning electron micrograph of an open stoma on a rose leaf

> It's a 'hypothesis' because it hasn't been completely proven but there is plenty of evidence to support it:
>
> - sucrose moves through phloem 10 000 times faster than it could by diffusion alone
>
> - concentration of sap is higher near the source and there are pressure differences between the source and sink
>
> - ringing experiments and radioactive tracers have shown organic solutes, like sucrose, can move through phloem in either direction, which is difficult to explain by a simple passive process like transpiration
>
> - there's loads of ATP in sieve elements and mitochondria in companion cells, suggesting some sort of active transport
>
> - phloem seems to metabolise fastest when translocation is rapid
>
> - killing phloem with metabolic poisons that stop active transport also stops translocation

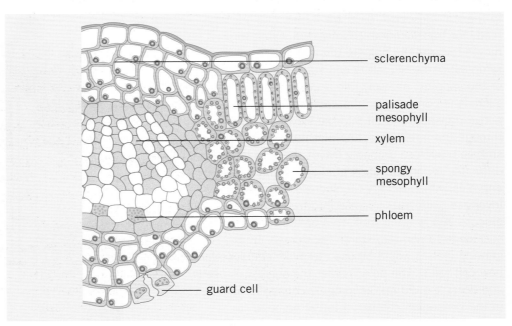

sclerenchyma

palisade mesophyll

xylem

spongy mesophyll

phloem

guard cell

Annotated section through a dicotyledonous mesophyte leaf

The upper epidermis is a single layer of cells without chloroplasts, coated with a waterproof waxy cuticle. There aren't many stomata as they would lose too much water on the hot, upper surface of the leaf, facing the sun.

Light goes through the upper epidermis into the palisade mesophyll layer, where most photosynthesis happens. The tall, slim **palisade cells**, packed with chloroplasts, trap as much light energy as possible, even if rays hit the leaf at an angle. Air spaces between the cells allow gas exchange.

The spongy layer is less important for photosynthesis, but has bigger air gaps (hence the sponginess) to allow easy diffusion of gases between the stomata and the palisade layer. The lower epidermis lacks a waxy cuticle and has most of the stomata, which control gas exchange, in and out of the leaf. During the day, carbon dioxide will be diffusing into the leaf for photosynthesis. The leaf makes oxygen as a by-product of photosynthesis faster than its cells use it for respiration, and the excess oxygen diffuses out. At night, photosynthesis stops. Oxygen diffuses into the leaf for respiration, and waste carbon dioxide diffuses out. The production and consumption of these gases is enough to maintain a diffusion gradient with the outside atmosphere so long as the stomata stay open.

Vascular bundles contain phloem and xylem and pass down the **midrib** of the leaf. This vein divides into finer and finer branches, forming a network through the whole leaf. The xylem brings water and mineral ions for photosynthesis, and the phloem takes away the dissolved carbohydrates made by photosynthesis.

How do stomata open and close?

Plants save water by closing their **stomata** (singular: stoma). They normally do this at night, when they don't need carbon dioxide for photosynthesis. Enough oxygen can diffuse in through the lower epidermis to allow respiration to continue.

A stoma is actually the empty gap between a pair of guard cells. It is the guard cells that open and close the stoma. They have cellulose microfibrils in the inner walls, lining the stoma, which stop these walls stretching. The pair of **guard cells** are joined at the ends. As a result of these features, when the guard cells fill up with water and become turgid, they bow outwards – opening up the stoma. When they lose turgidity, the stoma closes again.

Plants have an emergency override system to close up the stomata during the day if they are losing water through transpiration faster than they can replace it. This anti-wilting measure involves a plant hormone called **abscissic acid** (ABA), which causes K^+ ions to be quickly pumped back out of the guard cells into the adjacent cells, even in daylight. The stomata close, conserving water. Photosynthesis grinds to a halt due to lack of carbon dioxide, but at least the plant survives!

Macrophotograph of part of a dicotyledonous Maple leaf (Acer sp.), showing the branching veins

light

chloroplasts provide energy, to pump potassium, by photosynthesis

potassium pumps switched on by light

potassium ions build up in the guard cell so water moves into cell, down a water potential gradient

K+   K+

water   water

guard cells

adjacent cell, loses potassium ions and water to the guard cell

Potassium pump mechanism of stomatal opening

Light striking the guard cells (or low carbon dioxide levels) switches on **potassium pumps**, which actively transport K^+ ions into the guard cells from the adjacent cells. This gives the guard cells a more negative water potential than the adjacent cells, so they take in water and become turgid, opening the stoma. In the dark, the potassium pumps stop and K^+ ions diffuse back out of the guard cells. Water leaves the guard cells and the stoma closes.

Summary

Large organisms require specialised systems to exchange materials with the external environment and to transport them around the cells.

Lungs provide a large area for gas exchange and the circulatory system delivers oxygen to cells and takes away carbon dioxide and water. Feedback mechanisms, involving chemical and nerve signals, regulate breathing and heart rate in response to changing demand. Foetal haemoglobin, haemoglobin and myoglobin are adapted to transport oxygen in different circumstances. Anaerobic respiration allows muscles to continue working when there is an insufficient supply of oxygen.

The digestive system uses mechanical and chemical processes to break down large, insoluble food molecules into small, soluble molecules that are absorbed and transported in the blood to supply cells with fuel and raw materials.

Plants absorb carbon dioxide for photosynthesis but lose water as a result. Transpiration is the movement of water through a plant and occurs in the xylem tissue, which also transports dissolved inorganic ions. Phloem transports organic substances both up and down the stem – a process called translocation.

Practice questions

1 The diagram shows the dissociation curve of haemoglobin. Use the curve to:

  a explain how haemoglobin is adapted to carry oxygen

  b illustrate the Bohr effect by drawing a second graph

  c show how foetal haemoglobin works, by drawing a third graph.

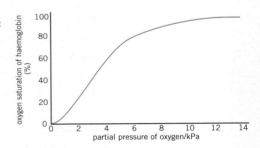

2a Describe four features of the lungs that make them efficient at gas exchange.

  b Explain how pressure changes in the thorax are caused in one breathing cycle.

  c Describe the role of the nervous system in the ventilation rate during exercise.

3a Explain how an increase in wind speed and temperature and a decrease in humidity make the water potential of air outside a leaf more negative.

  b Describe the difference between the apoplast and symplast pathways of water.

  c Suggest three adaptations that allow xerophytes to survive in desert conditions.

Key skills

You could satisfy all six key skills by collecting and analysing health statistics, for example diets and possible links to heart disease.

Can you design and carry out experiments to investigate the affects of exercise on the body or show how much exercise needs to be taken for significant improvement in aerobic fitness? You can get credit in all six key skills.

What factors affect the rate of transpiration? Design and carry out experiments in the right way and all six key skills are yours.

Ecosystems

In this section you will learn:

☞ what ecosystems are

☞ about the different ways organisms obtain nourishment

☞ how energy flows through an ecosystem

☞ how materials are reused in the carbon, nitrogen and water cycles

☞ about the impact of human activity on the environment.

Ecology is a branch of biology that is concerned with the interactions between organisms and the living and non-living components of their environment. It is big-picture stuff, studying relationships between populations of living things and their surroundings, rather than considering individual organisms in isolation. Studying the dynamics of whole ecosystems is a relatively recent area of science. Even more recent are attempts to understand how ecosystems fit together and how the whole balance of the Earth's biosphere works.

Understanding what makes ecosystems tick is becoming vitally important. There is a broad consensus amongst scientists (and governments, to varying extents) that human activities are causing fundamental shifts in the global ecology. In a relatively short space of time, an exploding human population has had a dramatic impact on the environment. Our accelerating consumption of limited resources is literally changing the face of the planet. It isn't only the effects of pollution that are a cause for concern, although global warming, damage to the ozone layer, acid rain, and reduced air and water quality are direct results of human industry. Loss of habitats, fragmentation of habitats, the effects of introduced species and over-harvesting are also taking their toll on the world's biodiversity. Steps to contain, and perhaps even reverse, these problems include increased use of renewable energy sources, improved energy efficiency, sustainable farming practices and more reuse and recycling of manufactured materials. These are areas where the developed countries, having been the biggest contributors to the problem, could be leading the way in providing real solutions.

Life on Earth is a complicated and fascinating phenomenon, and studying it is a complicated business. The underlying principles, however, are not difficult to grasp. This section builds upon work you did for GCSE Biology, reviewing the basics and then taking the concepts a little further. Even if you choose not to study Biology beyond AS level, an understanding of the ideas in this final section will equip you to make an informed contribution to the debate about sustainable development.

Nutrition and feeding strategies

Autotrophic and heterotrophic nutrition

You'll remember from GCSE that food chains begin with producers, usually green plants. Green plants make their own food, building up complex organic molecules, such as glucose, from simple inorganic ones, such as water and carbon dioxide. This is called **autotrophic nutrition** – literally translated as 'self-feeding'. Organisms that do this are called **autotrophs**. Using energy in the form of sunlight, the chloroplasts in leaves reassemble the atoms in carbon dioxide and water into energy-rich carbohydrate. Photosynthesis is how energy gets into the food chain:

$$6CO_2 \quad + \quad 6H_2O \quad \longrightarrow \quad C_6H_{12}O_6 \quad + \quad 6O_2$$

$$\text{carbon dioxide} + \text{water} \quad \longrightarrow \quad \text{glucose} \quad + \quad \text{oxygen}$$

It's a bit like charging up a molecular battery. The glucose on the right of the equation has a lot of chemical energy stored in its bonds, much more than the carbon dioxide and water on the left-hand side. This extra energy comes from light absorbed by the chlorophyll. Everything else in the food chain needs to eat either plants or other animals, breaking down complex organic molecules by digestion and resynthesising them into whatever they need. This is **heterotrophic nutrition** – translated as 'feeding on others', and organisms that do it are **heterotrophs**.

So, producers are autotrophs and consumers (including parasites and saprophytes) are heterotrophs. Heterotrophs rely on the chemical energy stored up by autotrophs. In a sense we are all solar-powered.

Strictly speaking, green plants are **photo-autotrophs**, as are some bacteria. There are other kinds of bacteria that don't get their energy from light, but instead from chemical reactions, and these bacteria are called **chemautotrophs**. Some are very important, such as those involved in the nitrogen cycle.

Holozoic nutrition

This is the kind of nutrition that animals, such as humans, go in for: feeding on organic matter from the bodies of other organisms. It involves four steps:

1 **ingestion** – taking food in

2 **digestion** – breaking down big, insoluble molecules into small, soluble ones

3 **absorption** of the soluble products of digestion into the body

4 **assimilation** – building them back up into the necessary complex molecules.

> Nutrition is such a fundamental process that it's no surprise to find that animals have evolved all sorts of adaptations to help them make the most of their food. Being a herbivore presents a particular challenge as plant cells are encased in walls made of cellulose, a particularly tough carbohydrate, which is hard to digest.

GURU TIP

You need to know the equations for both photosynthesis and respiration off by heart. This isn't hard, as one is the mirror image of the other.

Food is necessary for 3 reasons:

- energy to power all the reactions in cells and to release heat

- raw materials for growth, repair and development of cells and tissues

- vitamins and minerals required to keep all the body's biochemical reactions running smoothly.

Ruminants

Cows and sheep rely on bacteria to digest cellulose.

- First, they snip the grass with broad **incisors** at the front of their mouths
- Next, they grind up the grass with large, ridged **molars**. These teeth keep growing as they are continually getting ground down.
- The grass is swallowed and enters one of four stomach chambers. It's mixed with mucus and is formed into balls of 'cud', which are regurgitated into the mouth for a second chew, before being swallowed into the rumen (hence ruminant).
- The rumen is a big, anaerobic fermenting chamber. Here, vast colonies of bacteria mixed with saliva set to work on the grass, secreting cellulases, which break down the cellulose. A lot of waste gas is produced (mainly carbon dioxide and methane), which escapes from both ends of the animal.
- Eventually the cellulose is broken down and digestion can proceed more or less as normal through two more stomachs and the rest of the gut. The bacteria use some of the energy released from the grass, so it's not terribly efficient and ruminants need to spend all day eating, chewing the cud and passing wind.

GURU TIP
Ecology contains lots of science jargon. Do yourself a favour and learn the words in bold mean and what they mean (they're in the glossary at the back).

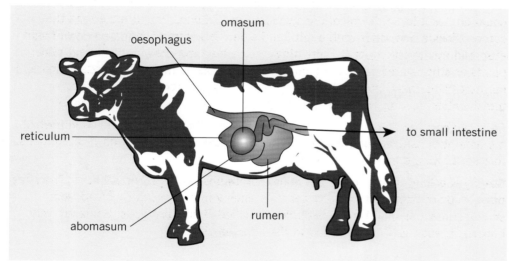

reticulum

oesophagus

omasum

to small intestine

abomasum

rumen

Cows have four stomach chambers

Turning grass into steak is quite tricky. As well as the cellulose problem, a diet of grass is low in protein, but again bacteria come to the rescue. **Chemautotrophs** can synthesise proteins from ammonia in the rumen. The proteins are ingested by single-celled **protocysts**, which pass along the gut and are digested by the cow, providing amino acids for the cow to build into proteins, as well as all the vitamin B$_{12}$ it needs. In return, the bacteria get a steady supply of food, good chemical conditions (pH and moisture), physical accommodation and warmth.

Carnivores

Carnivores have an easy time of it, by comparison. Eating the flesh of another animal is highly nutritious, and proteins from animals contain roughly the right proportions of all the necessary amino acids. The only difficulty is getting hold of prey, which will be doing its level best to avoid getting eaten. Once caught, lone hunters need to wolf down their prey quickly, before any company turns up. Some carnivores hunt in groups and have developed complicated social behaviour. Carnivorous mammals, such as lions and dogs, have distinctive long, sharp canine teeth for gripping; incisors for slicing and pointed molars for cutting. In general, carnivores have keen senses and quick brains. They need to be fast, too!

It seems likely that our distant ancestors had the ability to digest cellulose. Herbivores, such as rabbits, have a large appendix attached to their intestine, which contains cellulose-digesting bacteria.

Ecosystems

Nutrition and feeding strategies

Herbivores	Carnivores
Long digestive tract, often with a modified stomach with several chambers, a large **caecum** and **appendix**.	Relatively short digestive tract with a simpler stomach and smaller appendix.
Teeth keep growing to make up for continuous wear and tear.	Teeth stop growing when animal reaches adulthood.
Ridged molars for grinding.	Pointed molars for cutting.
Small canines.	Big, sharp canines for gripping.
Sharp, wide incisors for snipping. Upper incisors sometimes missing.	Thin, sharp incisors which act in pairs for biting.
Jaw moves from side to side to grind up plant material.	Jaw moves mainly up and down for cutting and chewing flesh and skin.

Comparing adaptations of herbivores and carnivores

Notice the way that the design of the whole oganism is shaped by the type of food it eats. Evolution, through natural selection, has produced organisms that are experts in their particular feeding strategy. This includes omnivorous feeders, such as rats, who thrive by eating just about anything they can find.

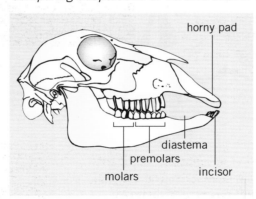
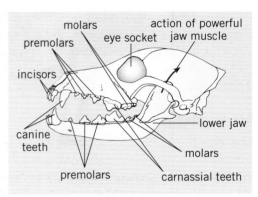

Comparing the adaptations in a sheep skull and a dog skull

Saprobiontic nutrition

Other important heterotrophs are the saprobionts. These are also called saprotrophs, or saprophytes and they feed on dead and decaying material. The group includes **decomposers** and **detritivores** that break down the dead remains of plants and animals, releasing the elements in the organic compounds of their bodies to be recycled. Some are economically important, such as the bacterium *Lactobacillus* used to make yoghurt and the fungus, *Serpula*, that causes dry rot.

Rhizopus is the mould that grows on bread if you leave it lying around. A typical example of a saprobiontic decomposer, it has no internal digestive system, relying instead on **extracellular digestion**. It sends thread-like hyphae deep into its food, secreting powerful cellulase and amylase enzymes, which **hydrolyse** the carbohydrates in the bread into simple sugars. These are absorbed through permeable cell walls. The hyphae have a big surface area and thin structure for diffusion and they grow fast under the right conditions.

Bread covered in Neurospora mould

Parasitic nutrition

Parasites are metabolically dependant on a host for most of their life cycle. They live on, or inside the host, feeding directly from it without even waiting until it's dead. Unlike the mutualistic relationship between ruminants and their gut bacteria, parasites have an uneven relationship with their hosts. The host receives nothing and often the parasite actually harms it to some extent. The most successful ones are careful about how much damage they cause, as a dead host is not much use.

The tapeworm, *Taenia*, is a common parasite of vertebrates, including domestic animals, farm animals and humans. It is brilliantly adapted to a parasitic lifestyle. Ranging in length from 1mm to several metres, its **scolex** lacks sense organs or a mouth. Instead, it uses hooks and suckers to attach itself to the wall of the host's small intestine. There it stays and grows, bathed in pre-digested food, absorbing nutrients straight through its body wall. It doesn't have a digestive system, although it does secrete some enzymes to speed up extracellular digestion outside its body wall, effectively competing with the host for food. The flat shape and thin, highly folded body wall gives it a large surface area and it has lots of mitochondria to power the active transport of food molecules that it can't get by diffusion.

Like many parasites, the tapeworm has a complicated and pretty disgusting life cycle. Getting transferred from one host to the next presents quite a challenge for a parasite and the solutions they have evolved makes parasitology a fascinating area of biology.

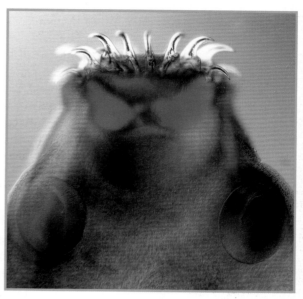

Light micrograph of a Taenia scolex

Mutualistic nutrition

This type of nutrition is called **symbiosis** (two organisms live in partnership, from which both gain some kind of advantage). Ruminants have mutualistic cellulose-digesting bacteria in their guts. The bacteria get warmth, food and shelter and the ruminant has help digesting cellulose. They have come to rely upon one another.

It's not just animals that strike up such alliances with micro-organisms. You might remember from GCSE that leguminous plants, such as peas, beans and clover have a neat trick for getting hold of nitrogen. They use **root nodules** full of **nitrogen-fixing bacteria**, such as in *Rhizobium*. These bacteria get into the root hairs of leguminous plants, where they stimulate the cortex cells to multiply into swellings that plumb directly into the root's vascular tissue. These nodules act as fertiliser factories for the host plant. *Rhizobium* converts atmospheric nitrogen gas (useless to the plant) into nitrates that the plant can use to build proteins. Farmers routinely 'infect' seeds with *Rhizobium* to make sure that they develop plenty of nodules, especially if they want crops to grow in poor soil. Genetic engineers are optimistic about the prospects of isolating genes that allow bacteria to fix nitrogen and inserting them into crops, such as wheat. Imagine the economic and environmental benefits of cereal crops that make their own fertiliser, literally out of thin air!

Root nodules on a pea plant

Ecosystems

Energy flow through ecosystems

Ecology is all about studying ecosystems, yet an 'ecosystem' is actually quite hard to define. The following terms are useful ones to get straight in your mind.

- The **biosphere** is the complete set of all the ecosystems on Earth. It includes all living organisms and the non-living part of the environment that they interact with. It involves the land, seas and the lower part of the atmosphere. If you're alive, or something that comes into contact with living things, you're part of the biosphere.
- An **ecosystem** is a more or less self-contained ecological unit. It includes all the organisms living in a particular area, together with their non-living surroundings. Ecologists call these the **biotic** (living) and **abiotic** (non-living) parts of an ecosystem. A pond is a good example of an ecosystem, containing a complex food web and abiotic factors, such as dissolved oxygen, water, light, rocks and so on. Sometimes it will interact with other ecosystems: birds carry seeds or frogs colonise another pond. Ecosystems can also be very big – the Sahara desert, for example.
- A **habitat** is the area within an ecosystem where an organism lives.
- A **population** is the number of individuals of a particular species in an ecosystem.
- A **community** is the total of all the populations living in an ecosystem.
- A **niche** describes the way an organism fits into its environment – not just where it lives, but how it goes about making a living.
- **Producers** are autotrophs that make their own food, for example, green plants are at the start of every food chain.
- **Consumers** are heterotrophs that make up the rest of the food chain.
- **Decomposers** are saprobionts that break down dead organic material, so nutrients can be recycled within an ecosystem. Most decomposers are bacteria and fungi.

Food chains and food webs

The simplest way to show who eats whom in an ecosystem is by a food chain.

rose plant → aphids → ladybirds → blue tits → hawk

The arrows are showing us how energy flows through the food chain. Each link in the chain is called a **trophic level**, and each level have a general name.

producer → 1st consumer → 2nd consumer → 3rd consumer → 4th consumer

1st, 2nd, 3rd and 4th consumers are also called primary, secondary, tertiary and quaternary consumers. The 1st consumer will be a herbivore, and the other consumers will be carnivores but the carnivores also get numbered.

producer → herbivore → 1st carnivore → 2nd carnivore → top carnivore

Notice how the 1st carnivore is actually the 2nd consumer. The problem with food chains is that they tell you nothing about the amount of organisms at each trophic level and it's unlikely that blue tits, for example, will only eat ladybirds. To give a fuller picture of the feeding relationships, you need to link several food chains together into a food web. If you're ever unsure what's going on, follow the arrows. These always show which way the energy flows, and each one represents a move up a trophic level. Here is a food web representing an Arctic food system. Notice how humans, who have no natural predators, are at the top of food chains and food webs.

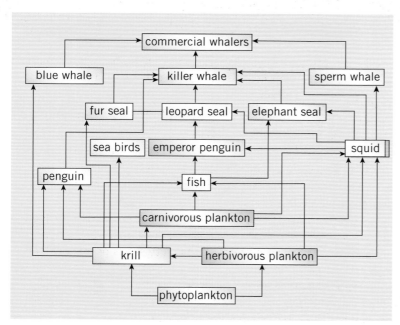

Antarctic food web

Pyramids of number and biomass

A food web gives a much better impression of what's happening but it is still only a **qualitative** description. In other words, there are no numbers. To show the relative numbers of individuals at each trophic level, you need a **pyramid of numbers**. This gives you **quantitative** information about a particular food chain.

The width (or sometimes the area) of the bar at each trophic level is proportional to the number of organisms. Here's one for the example food chain:

hawk

blue tits

ladybirds

aphids

rose plants

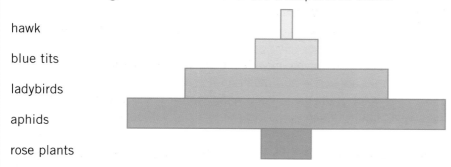

Notice how it isn't a pyramid shape at all. This can easily happen when some organisms in a food chain are a lot bigger than others. A single rose bush, for instance, could support many thousands of tiny aphids.

Ecologists get round this by using a **pyramid of biomass**. Instead of showing the number of individual organisms at each trophic level, this shows the total mass of all the individual organisms at each trophic level. Sometimes it is based on the 'wet' mass, as if you took all the individuals and simply weighed them. Other times 'dry' mass is used, which involves taking a sample of organisms from each level and heating them in an oven to remove all the water before weighing. This compensates for the fact that some organisms (for example,. plants covered with fleshy fruits) are naturally soggier than others, which would otherwise skew the results.

Ecosystems

KEY SKILLS
C3, N3, IT3, WO3, LP3, PS3

The important thing about a pyramid of biomass is that it is always a proper pyramid, with a wide base and each trophic level being narrower than the one below. It has to be, as you can't create mass out of nothing, so the total mass of organisms at a particular trophic level must be less than the total mass of the organisms they feed on. Here's a pyramid of biomass, for example:

hawk

blue tits

ladybirds

aphids

rose plants

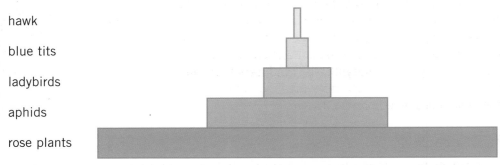

Notice the proper pyramid shape. Ecologists need all these ways of representing aspects of an ecosystem, as each method gives different information.

Energy transfer between trophic levels

You've probably noticed that food chains rarely go further than a quaternary consumer and are often shorter. There is a good reason for this, and it's all to do with energy. Exam questions often ask you to work out the percentage of energy that gets transferred from one trophic level to the next. The answer is usually not a lot, typically under 30% and often much less. So where has the rest of the energy gone? There are four main ways this energy gets lost: respiration, digestion, evasion and excretion.

1 **Respiration** - plants and animals use up most of the energy they get from photosynthesis or food, just staying alive. Only a small amount ends up as new biomass. Animals that are active and maintain a constant body temperature, such as mammals, need a lot of energy to keep warm, and for movement. Think how many kilojoules an adult human can consume in a day without actually putting on any weight.

2 **Digestion** is nowhere near 100% efficient. Many parts of plants and animals can't be digested at all – humans can't digest plant cellulose or animal bones, for example. Herbivores tend to eat the shoots, leaving the roots behind. Not all the digested food gets absorbed as some passes out as faeces. Plant protein contains different proportions of amino acids to animal protein so transfer between the two is less efficient than between animals. Also, don't forget that the process of digestion itself, requires energy.

3 **Not being eaten in the first place** – only a proportion of organisms at a particular trophic level are actually consumed by members of the next trophic level. Many die and decompose instead, removing energy from the food chain.

4 **Excretion** – getting rid of the waste products of metabolism requires more energy.

As energy is transferred up a food chain, most of it is lost along the way. Mainly it's lost as heat, which dissipates into the environment. Only a small proportion actually ends up being used to build new tissue available for the next trophic level. Food chains are generally quite short because they run out of usable energy.

Pyramids of energy

If the arrows in a food chain are showing the transfer of energy, why not draw a pyramid where each bar shows the amount of energy locked up in each trophic level? Here is what a pyramid of energy might look like for the Antarctic food chain (energy is measured in kJ):

This shows 10% of the energy available at each trophic level actually making it into the next (and this is a fairly efficient food chain compared to most). Only some light energy striking a plant gets absorbed by the chlorophyll. Of this energy, only a proportion is transferred into chemical energy by photosynthesis. Producers that can transfer just 4% of the light reaching them into new growth are considered pretty efficient.

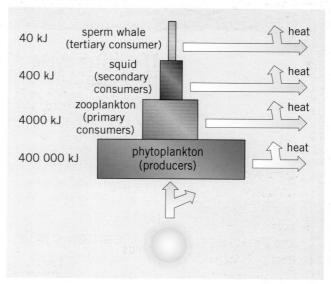

Pyramid of energy for an Antarctic food chain

Productivity in ecosystems

The **productivity** of the organisms dictates how elaborate a food chain can be. The rate at which producers absorb light and use it to build biomass is called the **gross primary production** of an ecosystem. Not all of this energy is available for the first consumers as the producers need some for respiration. Once this amount has been subtracted, the balance is the ecosytem's **net primary production** (energy available to consumers).

Productivity is measured in 'kilojoules per square metre per year' ($kJ\ m^{-2}\ yr^{-1}$) and tells us how much light is transferred into chemical energy in new plant tissue. Different ecosystems vary in primary productivity. Deserts, for example, have a lot of light energy but not enough water for plants to grow easily – they have a low primary productivity, so few consumer species can be supported. Tropical rainforests, on the other hand, have plenty of light and water and the primary productivity is very high. This supports a huge, intricate food web containing more species than we can name. Tropical rainforests are said to have a high biodiversity, which is one of the reasons why there is so much concern about their present rate of destruction.

Improving productivity

Farmers and agricultural scientists need to maximise their crop productivity. They measure crop yield in 'tonnes per hectare per year'. There are four main ways you can boost yield, and all are widely used throughout the developed world.

1 **Improve the abiotic conditions** to increase the rate of photosynthesis, for instance, tomatoes are grown in greenhouses all year and crops are irrigated.
2 **Add fertilisers** to increase growth by supplying nitrogen, phosphorous and potassium, which tend to get removed from the soil as crops are harvested.
3 **Use selective breeding** to produce high-yield crops by crossing parent plants with desirable characteristics, such as disease resistance. Genetic modification is used increasingly, to produce new varieties with useful characteristics.
4 **Use pesticides and herbicides** to stop valuable crops being eaten or destroyed by disease and reduce competition for light, space and minerals from weeds.

Farming in Britain is intensive – a lot of energy goes into each hectare of crop in addition to sunlight, in the form of chemical fertilisers, pesticides and herbicides, as well as the energy required to run the farm machinery. Most of this additional energy comes from oil products, which the farming industry is very reliant upon.

Recycling of materials

The water cycle

The sun evaporates water from oceans and lakes and this vapour rises and condenses into clouds, which return water to the ground by **precipitation**. The precipitation that falls on land either evaporates or returns to the oceans and a small amount drains into underground aquifers. Metabolic processes have a limited effect on the global cycle but it can be significant in local areas, such as rainforests. Plant roots are important as they hold soil together and improve drainage. Clearance of plants, either through natural effects or human intervention has a drastic effect on the local water cycle, leading to mudslides, desertification and flooding.

The carbon cycle

This represents a balance between three processes: photosynthesis, respiration and combustion. Try and remember the following important points:

1 Photosynthesis uses CO_2 by absorbing it and converts it into organic compounds.
2 Carbon, in the form of these organic compounds, passes along the food chain and is released as CO_2 during respiration. Dead, organic remains are broken down by decomposers, which release carbon back into the atmosphere.
3 **Phytoplankton** are an important carbon 'sink'. They absorb CO_2 by photosynthesis and are then eaten by **zooplankton**, which use some carbon to make their shells. When they die, they sink to the bottom, removing carbon from the cycle for good. Have a look at the white cliffs of Dover to get an idea of the scale of this process.
4 Some CO_2 dissolves in the oceans, or forms hydrogen carbonate ions (HCO_3^-).
5 Millions of years ago, forests acted as carbon sinks. Their fossilised remains became coal, oil and gas, locking away hydrocarbons. Humans burn these fossil fuels, increasing the proportion of carbon existing as CO_2 in the atmosphere.

Remember: combustion of any kind makes carbon dioxide and water. We're burning fossil fuels much faster than new deposits are being laid down.

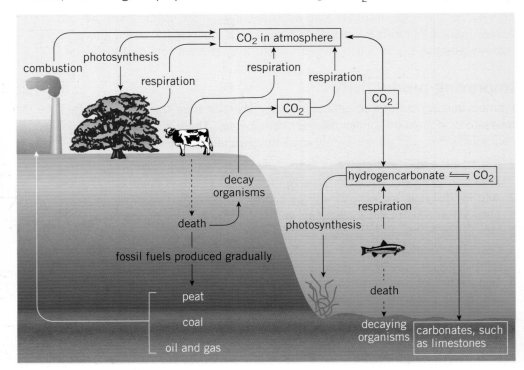

The carbon cycle

AS Guru™ Biology

The nitrogen cycle

This is a bit more complicated, but take it step-by-step and you'll get the hang of it. All organisms need nitrogen to make amino acids (the building blocks of protein). Nitrogen circulates around ecosystems, as plants and animals are digested and decompose. Nitrogen gas is made of pairs of atoms (the chemical formula is N_2), held together by a strong triple covalent bond, which makes it very unreactive (inert). It has to be converted into a more reactive, soluble compound, such as ammonia (NH_3) or nitrate (NO_3^-) that can be absorbed by plants before entering the food chain – a process called nitrogen fixation. Fixation happens in three ways:

- lightning converts atmospheric nitrogen to nitrogen oxides, which dissolve in rain
- the **Haber process** is an industrial method of making ammonia out of nitrogen and hydrogen, which is then turned into ammonium nitrate fertilizer
- bacteria can fix nitrogen, including *Rhizobium* (mainly found in root nodules).

Once plants have absorbed nitrogen it passes up the food chain. Nitrogen returns to the environment from living things in two ways:

- excess amino acids are broken down by **deamination** in the livers of animals, and the nitrogen ends up in urea, which is excreted as urine
- decomposers (mainly fungi and bacteria) break down amino acids in dead plants and animals, releasing ammonia. **Nitrifying bacteria**, such as *Nitrosomonas* and *Nitrobacter* quickly turn this into nitrite ions (NO_2^-) and then nitrate ions (NO_3^-).

Human activity can disrupt the nitrogen cycle. Nitrate-rich run-off from crop fertilizers and livestock gets into rivers and lakes where it causes **eutrophication**. The nitrates encourage blooms of algae to grow, which then die and sink. Aerobic bacteria decompose the dead algae, using up most of the dissolved oxygen in the water. The lower oxygen level kills many invertebrates, starving the vertebrates higher up the food chain of food and oxygen, thus reducing biodiversity. Raw sewage discharged into rivers or the sea has a similar effect, for the same reason.

GURU TIP
The key thing to grasp is that, although the air is 79% nitrogen, living organisms can't get hold of it directly.

GURU TIP
Denitrifying bacteria, (for example, *Pseudomonas* and *Thiobacillus*) are chemautotrophs, which turn nitrates into nitrogen gas.

The nitrogen cycle

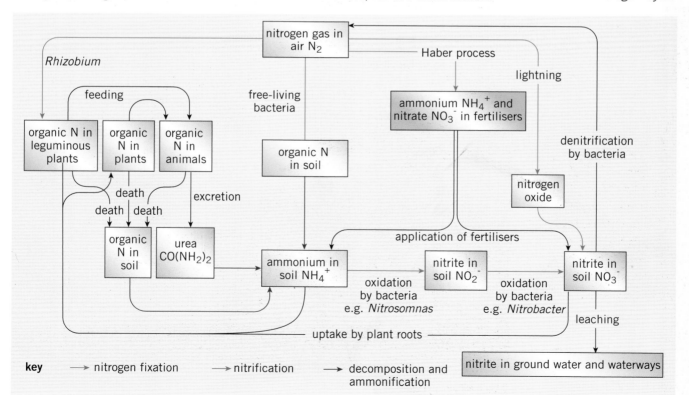

Ecosystems

Human effects on the environment

KEY SKILLS
C3, WO3, LP3

GURU TIP

The Greenhouse Effect and the hole in the ozone layer are different phenomena, although CFCs actually contribute to both. Ozone (O_3) depletion by CFCs is damaging the ozone layer high in the atmosphere, which acts as a shield against harmful ultra-violet radiation from the Sun.

Acid rain

The widespread combustion of fossil fuels releases huge amounts of carbon dioxide into the air. However, it also releases acidic gases, such as sulphur dioxide and nitrogen oxides, which dissolve in the rain and form sulphuric acid and nitric acid. As well as damaging buildings, rain with a lower pH (more acidic) has harmful effects on living organisms, mainly by altering the availability of dissolved minerals:

- plants lose leaves and die, removing producers from the ecosystems
- essential mineral ions become less soluble and plants can't absorb them
- phosphate ions (PO_4^-) bind to clay particles, so plants can't use them
- aluminium ions (Al^{3+}) can reach poisonous levels, which can kill fish in lakes.

Lakes and woodland in northern Europe have been acidified, largely, by air pollution from the coal-fired power stations in Britain. Acidified lakes often look beautiful with sparkling blue water, but the water is clear because there is so little living in it. Lakes with low levels of available minerals are **oligotrophic** – the opposite problem to having too much nitrate and phosphate, which leads to eutrophication.

The Greenhouse Effect

The Greenhouse Effect is essential for life on Earth. Without it, the planet would be frozen and incapable of supporting life. However, the Greenhouse Effect seems to be increasing, leading to global warming. Here's a reminder of how it works:

- electromagnetic energy of different wavelengths reaches the Earth from the Sun, and most of this radiant energy is let through the atmosphere to the ground
- the land and seas are generally dark in colour, maximising the absorption of visible light, which is re-emitted at longer, infra-red wavelengths (heat)
- gases in the atmosphere trap these wavelengths, preventing heat from escaping.

There are many greenhouse gases, such as water vapour, CO_2, methane, ozone, nitrogen oxides and chlorofluorocarbons (CFCs) but there is a relatively large amount of CO_2 in the atmosphere (around 0.03%), so it has the greatest effect. The concentration of CO_2 is increasing measurably, as a direct result of the human destruction of forests and burning of fossil fuels. There is mounting evidence to suggest that the resulting rise in global temperature is having a profound effect on weather patterns. Melting ice caps will cause the sea level to rise, leading to flooding. Changes in climate are proving very difficult to predict, but the general consensus is that continued global warming would do much more harm than good.

sun

atmosphere lets short wavelength radiation through

Greenhouse gases trap infra-red radiation

Greenhouse gases trap infra-red radiation

Deforestation

An area of rainforest about the size of England is being destroyed each year. This is a concern on a number of counts.

- Firstly, the destruction of habitats, and the resulting extinction of countless species, are irreversible. The phenomenal biodiversity of the rainforest ecosystem includes many species that have already provided useful drugs and there are doubtless many more waiting to be discovered.
- Secondly, cleared land suffers from rapid soil erosion and soon becomes infertile, once the trees have been removed. After a few years, yields fall and more forest has to be cleared.
- Finally, much wood is burned either as fuel or to clear land, releasing large amounts of carbon dioxide.

The economic pressures to cut down rainforest are very strong in some parts of the world. Tropical hardwoods, such as mahogany are valuable building materials, and developing countries need land for grazing cattle and growing cash crops, such as coffee. Reversing the present trend will be hard – at the present rate most of the world's rainforests will be gone by the time your children have grown up.

It is possible to manage forest resources in a sustainable way. This already happens with European pine forests, used to supply paper. As one area is felled, another is replanted to provide a continuous supply of wood. This is obviously much easier with fast-growing pine trees than with tropical hardwoods and these forests support nothing like the biodiversity of a tropical rainforest.

Tropical rainforest supports high biodiversity

Managing energy resources

Acid rain and global warming are largely a result of a dependence on burning fossil fuels for energy. Improving energy efficiency and finding renewable alternatives to coal, oil and gas are widely seen as the way to reduce pollution and avoid using up the limited supplies of fossil fuels that remain. Some alternatives to fossil fuels include:

- **Biogas** – mainly methane, like natural gas, but it is collected from rotting agricultural or domestic waste, rather than being a fossil fuel
- **Biomass** – using fast-growing crops as a fuel
- **Gasohol** – ferment sugar, using yeast, to make ethanol. Brazil has little in the way of oil reserves for making petrol, so most of their cars run on gasohol. The drawback is that land used for growing sugar cane is unavailable for growing food.

Nuclear power was originally thought to be able to generate electricity so cheaply that it would be too cheap to meter. However, despite the fact that it doesn't produce air pollution, the difficulties of managing radioactive waste and the cost of decommissioning old reactors has prevented nuclear power living up to earlier hopes. **Fusion power** seems as far away as ever, although research continues and some progress is being made with alternative ways of generating electricity using solar, wind or wave energy.

It remains to be seen whether safe technologies can be adopted, to reduce the current dependence on fossil fuels, in time to avoid lasting disruption to the world's ecosystems.

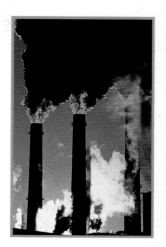

Coal-fired power station with yellow sulphurous smoke coming from the tall stacks

Ecosystems

Summary

The organisms within an ecosystem are interdependent – each organism has an effect on the rest of the community.

Food chains and webs are ways of describing the energy flow through an ecosystem. Pyramids of number, biomass and energy are alternative ways of illustrating feeding relationships. The primary productivity of an ecosystem is determined by the efficiency of photosynthesis, as most food chains start with photo-autotrophs. Energy is lost between trophic levels so there is a limit to how many links a food chain can sustain.

Carbon, nitrogen and water are recycled in ecosystems but the nitrogen cycle is particularly complex because plants cannot use atmospheric nitrogen directly. Combustion of fossil fuels is releasing an increasing amount of carbon dioxide that is probably contributing towards a rise in global temperature. There are growing signs that human activity is having a widespread and damaging impact on the environment.

Practice questions

1   The diagram shows a pyramid of numbers for a simple food chain.

hawk
blue tits
ladybirds
aphids
rose plants

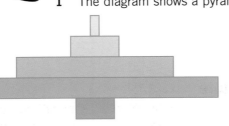

a Why is the bar representing the number of primary consumers wider than the one above?

b Identify an autotroph and a predator in the chain

c Explain three ways in which energy is lost between each trophic level of a food chain.

2a Big carnivores tend to live in small groups and have large territories in which to hunt. Suggest reasons for this.

b Suggest why birds and mammals lose more energy than other organisms.

c A recent study compared the productivity of mature tropical rain forest with intensively farmed cereal crops. The rainforest had the highest gross primary productivity but the crops had the highest net primary productivity. Can you account for the difference?

3a Why can't plants use atmospheric nitrogen to synthesise their amino acids?

b Describe two natural processes that fix atmospheric nitrogen.

c Explain how discharge of untreated sewage into a river can lead to eutrophication.

Key skills

Design and carry out experiments to show how energy is transferred through food chains and webs and investigate how efficient the process is, to get credit in all six of the key skill areas.

Are you able to demonstrate that improving the abiotic factors for a particular plant increases crop yield? Alternatively, can you show the affect of growing a particular species of plant in different soil conditions and quality, on the growth of the plant? Such projects would give you opportunities to demonstrate all key skills. Look through the checklist in the book's introduction to see what you need to do.

Coursework tips

Stay on course

- The proportion of your final mark awarded for coursework is 15%–30%, depending on which course you're following. Even 15% can have a drastic effect on your result. The difference between incomplete, low quality coursework and finished, good quality coursework can easily make the difference between one or even two grades.

Check the specs

- Know the skills being assessed and the marking schemes. These can be found in the exam specifications, although they may need translating into plain English! Find out what is expected from you, in terms of practical write-ups, drawings, teacher-assessed lab skills and an individual study or project.
- If your teacher hasn't supplied you with this information, ask for it. You don't have to buy a copy of the whole specification, as much of it won't apply to you.
- You can find out the coursework details by looking your exam body's web site. Download the specification for your course and then just print the bits you need:

  AQA: http://www.aqa.org.uk/

  EDEXCEL: http://www.edexcel.org.uk/

  OCR: http://www.ocr.org.uk/

KISS – Keep It Simple, Stupid!

- Don't bulk up your work with loads of pointless information. Stick to the point and target the marks. You'll gain credit for the quality of your work, not the quantity. Some exam bodies tell you how long your individual study report should be. Ask your teacher to show you some examples of good projects from previous students. This is perfectly acceptable and will give you a feel for what is required.

Get a grip on your variables

- OCR calls these 'key factors'. These are the things you change, measure or take account of. In planning, doing and interpreting results you need to have a clear idea of the point of your experiment. This is a common source of missed marks. Your investigation should start with an idea that you can test by experiment, also called a testable hypothesis. If you can't sum up your project in a sentence, you've got yourself in a muddle!

Use the right graph

- Plan to get quantitative data that will make a good graph (results that will provide real numbers so you have at least 5 points to plot).
- Choose the appropriate type of graph (line chart, pie chart, bar graph or histogram) to represent your results. There are clear rules for this – ask your teacher which graph is right for your data. If you use something unusual, such as kite diagrams, make sure you know what you are doing.
- Graphs must be labelled, with clearly plotted points and a line of best fit – make sure you include all of these to avoid losing easy marks.

TOP TIP

Coursework matters, especially if you find examinations difficult. If you follow these tips, and your teacher's guidance, just about anybody can produce good coursework and boost their marks. Think of it as an opportunity for success rather than a source of extra work and stress.

TOP TIP

Experiments give useful results when you change one thing and measure it's effect on something else to test your hypothesis. You get credit for mentioning variables that might affect your results but are beyond your control. Make sure you're on the right track by getting your plan checked by your teacher.

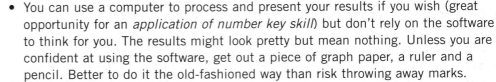

Coursework tips

- You can use a computer to process and present your results if you wish (great opportunity for an *application of number key skill*) but don't rely on the software to think for you. The results might look pretty but mean nothing. Unless you are confident at using the software, get out a piece of graph paper, a ruler and a pencil. Better to do it the old-fashioned way than risk throwing away marks.
- Find out what kind of graph your exam body likes. Do you need to show error bars? How do they prefer you to plot intercepts or points on your graph? Is there a mark or two going for calculating the gradient of a graph?
- Check the specs and ask the expert – your teacher.

Go steady on the stats

- You may need to do some statistical analysis of your results. The point of this is always the same: are your results telling you what you think they are? Make sure you're absolutely clear what you're testing for (your null hypothesis), don't try doing statistics just for the sake of it – you could miss out on marks.
- If you do use stats, check that you're using the right test for the kind of data you've collected before spending hours slaving over your calculator.

Present your work properly

- Don't throw marks away with sloppy spelling, punctuation and grammar. Use a word processor by all means but don't put all your faith in spelling and grammar checkers – get a trusted human being to check over a draft of your work.
- Keep tables, drawings, and diagrams large, clear and uncluttered. Avoid using colour unless you need it to make information stand out clearly (pie charts, for example). Use photographs with care – they're really only useful if they show something important that can't be illustrated by a drawing or diagram. Don't go mad on illustrated title pages or covers as it won't earn you extra marks.
- Hand in your pages in the right order, with your name on the front and held together with something like a simple treasury tab. Never put each sheet in a plastic sleeve – your teacher needs to be able to write notes on your work and this is will drive him or her insane!
- Make sure any unsupervised work you have done is clearly your own. All the exam bodies are very strict about this as the marks count directly towards your grade.
- If you've produced a really good piece of work, keep a copy in case it's sent away for moderation and not returned to school for ages. Biology coursework is a great addition to your key skills portfolio. You've done the work, so you deserve the credit!

Less is more

- You are expected to write in a scientific style, so keep to the point and write in the third person. Using appropriate specialist vocabulary shows that you know what you're on about and saves time and effort. It doesn't come naturally to many people, so be prepared to do a few draft versions before you get it right. Ask to see examples of good work and look at the style of language as you'll get credit for this.
- Before handing in your individual study, do a final check against the marking criteria to make sure that you've covered everything. For example, commenting on the reliability of evidence and listing sources of error. Don't miss easy marks by failing to include any bits and pieces that are specified, such as a list of references.

TOP TIP
Be clear in your mind why you are drawing a graph in the first place. If it doesn't serve a purpose, don't bother. One good graph that tells the story of your results puts marks in the bag.

TOP TIP
As with graphs, be very careful if you use a computer. Showing each step in your calculation helps you keep track of what you're doing and picks up marks.

TOP TIP
Some courses require an abstract at the beginning of your individual study. This is a short summary saying what your investigation was about and what the results showed. It's a good idea to include one anyway, as it will keep you on track and gives the person marking your work a quick overview.

Meet your deadlines

- Write down deadlines and organise your time in order to meet them.
- Don't stick your head in the sand! AS level isn't supposed to be easy – be prepared to put your back into your work. You'll have to organise your time much more than you're used to doing, if you're going to keep on top of things.
- Studying five AS levels is tough enough without making things even harder on yourself, so try not to fall behind. This will cut down the time your teacher has to check work and can give you feedback on areas for improvement. Knowing that you've got a big chunk of work overdue can be depressing and demoralising.
- Prioritise your work, and don't leave write-ups half finished. Stick at it until it's done and get it handed in, even if it's not absolutely perfect. The whole point about coursework is that you build up credit for the best parts of lots of different pieces of work, no single piece is worth a major stress attack.
- Track your progress (the exam boards provide a candidate record sheet). As you hand in work, you can note the date and see that you're making headway and as you progress through the course, you can spot any weak areas and do something about it.
- The beauty of coursework is that you get lots of chances to get things right and only your best marks in each skill area stick.
- Targeting your weakest marks for improvement is a very effective way to boost your final mark, as well as improving your scientific abilities.

Ask for help

- Act on the advice that's offered. Reading this shows that you are motivated enough to make an effort to succeed in your studies. You know that your exam grades are important, and that you will use some of the skills and knowledge that you are gaining now in your future life whether or not you study Biology any further.
- Producing good coursework involves quite a complex blend of skills and a fair degree of self-discipline. It's only natural that you'll get stuck from time to time. The trick is to have ways to get yourself unstuck. Your form tutor and subject teachers are actually on your side and have got the expertise and experience to help you out. Make the most of it!
- Bear in mind that getting specific help with a particular piece of coursework may affect your mark – your teacher will let you know if this is likely to happen. Even then, dropping a mark or two is a lot better than handing in some garbled nonsense that is hardly worth any marks at all.
- If you go on a residential field trip, get as much help as you can from the course tutor.
- Students in your group are another excellent source of help. Get together and talk about your work, outside of lessons. You're all in it together and there's a lot you can do to support one another. Sometimes the best teacher is someone who has only just learned how to do something themselves.

TOP TIP

Cut out any waffle. There is a strong temptation to pad out your work in the hope that you'll somehow pick up some extra marks. Resist! Marks are awarded against very clear criteria, and 'number of ink cartridges used' isn't one of them.

TOP TIP

If you do fall behind, ask your form tutor or teacher for advice. They can help you prioritise your work and give you practical tips on how to manage your time. Don't try to fob your teacher off with excuses for missing an important deadline. As time passes, you'll forget details or even what the piece of coursework was about and miss out important mark-earning points.

Answers

Section 1

1 a A rough ER   B cell wall   C nucleus   D nucleolus
      E chloroplast   F mitochondrion   G Golgi body   H vacuole

  b The cell wall and chloroplasts would be absent from an animal cell.

  c Resolution is the ability to distinguish between two points. If two points, lying close together, appear as one, they cannot be resolved. Electron microscopes can resolve closer points because electrons have a shorter wavelength than visible light.

2 Any four of:
    DNA – holds genes that control cell activity and replicates to produce new cells.
    Plasma membrane – a selectively permeable barrier that controls movement of substances in and out of the cell.
    Ribosomes – the organelles where proteins are assembled.
    Cell wall – gives physical strength to the cell.
    Cytoplasm – where the cell's metabolic activity takes place.

3 It doesn't matter if your spinning speeds or timings are a little different to those given below, so long as you include all the steps and they're in the right order: Put the liver in an isotonic solution and homogenise it by grinding; spin in a centrifuge at 700g, for 10 minutes; discard sediment; spin supernatant at 20 000g for 20 minutes; mitochondria should now be in the sediment.

Section 2

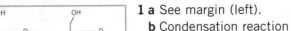

1 a See margin (left).

  b Condensation reaction

  c Glycosidic bond

2 a Primary structure describes the amino acid order in the polypeptide chain. Secondary structure describes the coiling or folding due to hydrogen bonding.

  b α-helix and β-pleated sheet

  c In globular proteins, the molecules form a spherical mass with a specific 3-D structure. Haemoglobin is a globular protein. In fibrous proteins, the molecules form long chains, often linked by hydrogen bonds. Collagen is a fibrous protein.

3 List as many properties as you can remember and then note down their biological implications. Assemble these points into continuous prose with a short introduction and conclusion. You'll get marks if you mention anything from this list:

Property	Biological significance
Hydrogen bonding/polar molecule.	Liquid at normal temperatures therefore provides a medium, in which organisms can live and can act as a transport medium.
Excellent solvent for ions and polar molecules.	Most reactions in living organisms take place in solution.
Thermal properties/high heat capacity/takes a lot of energy to raise the temperature of water	Large bodies of water provide a stable habitat; easier for organisms to maintain a steady body temperature; evaporation removes heat efficiently in panting and sweating; low freezing point means bodies of water and water in organisms less prone to freezing.
Ice is less dense than water.	Ice floats, insulating the water underneath so that large bodies of water are less likely to freeze solid.
High cohesion and surface tension.	Helps mass flow through vascular tissue in plants; high surface tension provides a habitat.

Section 3

1 a A hydrophilic phosphate group; B hydrophobic fatty acid chain

b Carrier and channel proteins transport materials across the membrane. Proteins can be receptors for molecules, such as hormones. Proteins are important in communication or connection between cells. Enzymes are also proteins.

c Hydrophilic phosphate groups face out, attracted to the water molecules of the aqueous medium surrounding the membrane. The hydrophobic fatty acid chains point in, forming the non-polar hydrophobic interior of the phospholipid bilayer.

2 a Molecules possess kinetic energy. In gases and liquids this makes them move randomly, so that they spread out within a given volume of space until they have reached a uniform concentration. This phenomenon is called diffusion.

b Choose three from the following:
  • thin membrane • large surface • steep concentration gradient • moistness.

c Facilitated diffusion is the movement of water-soluble molecules through specific intrinsic carrier protein molecules or channel protein molecules in the cell surface membrane. It does not require energy.

3 a Osmosis is a special case of diffusion as it only involves water and requires a selectively permeable membrane. Solute molecules are too large to pass through the selectively permeable membrane, causing a lower water potential on the solute side of the membrane and water diffuses across the membrane, down a water potential gradient. There is a net movement of water from a dilute solution to a more concentrated solution.

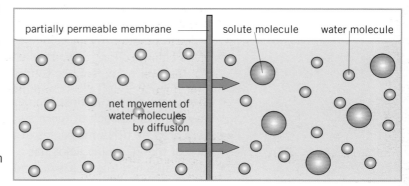

b The water potential of pure water is zero. The presence of solute molecules lowers the water potential. The more solute, the more negative (lower) water potential becomes, hence any solution will have a negative water potential.

c Hydrolysis of ATP to ADP + P.

Section 4

1 a See margin (right).

b As temperature increases, substrates and enzymes gain kinetic energy, increasing the frequency of interactions and hence the rate of reaction. Also, the higher temperature increases the vibration of the molecules, placing a strain on the structural bonds. When these bonds break, the enzyme changes shape, so the substrate no longer fits the active site, making the enzyme inactive. As more enzyme molecules become denatured in this way, the rate of the reaction slows. Increasing the temperature increases the rate of reaction to the optimum temperature, beyond which the rate falls as the enzyme molecules become denatured.

c Turnover rate is the number of substrate molecules that a single molecule of enzyme can act upon, in one minute.

2 a In photosynthesis, the light-independent stage happens straight after the light-dependent stage (not in the dark).

b CO_2 is used to convert ribulose biphosphate (RuBP) into glycerate-3-phosphate (GP) in the light-independent stage.

c In the stroma.

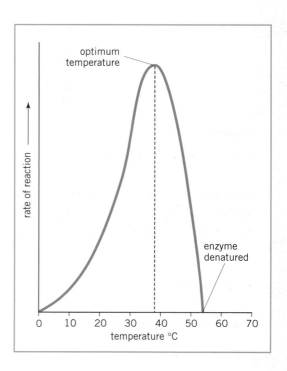

Answers

3 a Inhibitors slow down or stop enzyme activity. Competitive inhibitor molecules have a similar shape to the substrate molecule and therefore compete for the active site. Non-competitive inhibitors may be a completely different shape to the substrate. They bind to the enzyme away from the active site, but alter the shape of the active site so that it can no longer interact with the substrate.

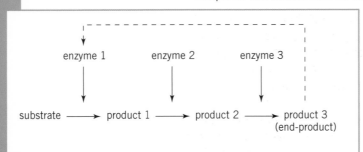

b An increase in the amount of product 3 inhibits enzyme 1, slowing down the production of products 2 and 3. Falling levels of product 3 allow enzyme 1 to function faster, boosting the production of products 2 and 3. The effect of end-product inhibition is to finely control the speed of the whole pathway.

c Mercury binds to enzymes permanently, irreversibly inactivating them.

Section 5

1 a mRNA: U A A U A C C G A C C U U A C C C U
DNA: A T T A T G G C T G G A A T G G G T

b 6 codons are shown on this section of mRNA.

c Check your answer against the description of translation on page 80. You should have mentioned each of these steps, in the correct order:
- there are free tRNA molecules in the ER, near the ribosomes
- each of the 20 amino acids has a tRNA molecule specific to it
- mRNA attaches to a ribosome and a tRNA (carrying the matching anticodon for the first codon on the mRNA) binds to it, bringing the correct amino acid with it
- the next tRNA binds to the adjacent mRNA codon, bringing another amino acid
- the amino acids are joined by a peptide bond, using energy from ATP
- the mRNA moves across the ribosome and the first tRNA detaches, ready to pick up another amino acid and be used again
- codons bind to tRNAs and keep doing so, until a complete polypeptide is made.

2 Deletion – a nucleotide is removed altogether.
Substitution – a nucleotide is exchanged for one with a different base.
Inversion – the order of a sequence of bases is reversed.

3 Check your answer with the information on page 81, making sure you included:
- isolation of the gene by a named method
- gene transfer, restriction endonucleases, ligases, sticky ends and plasmid vectors
- identification of recombinant bacteria using a genetic marker and replica plating
- cloning the bacteria
- expression of the insulin gene, collection and purification of insulin.

Section 6

1 a A protein filaments   B mitochondria   C head
D membrane   E nucleus   F acrosome

b Protein filaments contract, bending the tail. Mitochondria provide energy to move the tail, by respiration. The head contains a nucleus and an acrosome. A membrane covers the sperm. The nucleus has 23 chromosomes carrying the genetic information in the DNA. The acrosome makes enzymes that penetrate the egg.

c A sperm cell is haploid as it contains a single copy of the 23 chromosomes.

2 a Any four differences from the table on page 95.

  b Protandry (anthers ripen first), protogyny (stigmas ripen first) and dioecious plants (separate sex plants) are ways that flowering plants avoid self-pollination.

  c meiosis.

3 Compare your answer to the diagrams on pages 90–91. They should be simple and clearly show what happens to the chromosomes.

Section 7

1 a A bound oxygen molecule affects the affinity of a haemoglobin molecule for the next oxygen, so that haemoglobin has a high affinity for oxygen at high partial pressure (lungs) and low affinity at low partial pressure (tissues). See page 115.

  b Your second curve should be below and to the right of the first curve.

  c Your third curve should be above and to the left of the first curve.

2 a Moist (could mention surficant to prevent sticking); thin membrane; large surface area; ventilation and circulation of blood work together to maintain a steep concentration gradient. All these promote rapid gas exchange by diffusion.

2 b Check your answer against the description of a breathing cycle on page 105. You should relate contraction and relaxation of the diaphragm and intercostal muscles to changes in volume, and therefore pressure, in the thorax.

2 c Chemoreceptors in the medulla detect increases in blood CO_2 and H^+ ions. They stimulate inspiratory cells to send impulses to the intercostal muscles and diaphragm; breathing becomes faster and deeper. Chemoreceptors in the aortic and carotid bodies check the blood. If the concentration of CO_2 and H^+ ions is still high, they signal the medulla to send more frequent impulses to the intercostal muscles and diaphragm, causing even faster, deeper breathing.

3 a Discuss the number of water molecules per unit volume of air and how wind, temperature and humidity affect the water potential in the air, next to the leaf.

  b Apoplast pathway – water goes through cell walls and intercellular spaces. Symplast pathway – water goes through the cytoplasm of the cells.

  c Any three from the list on page 126.

Section 8

1 a Aphids are tiny and one plant can support thousands of primary consumers.

  b Rose plants are autotrophs. Ladybirds, blue tits and hawks are all predators.

  c You should explain energy loss by three of: inefficiency of photosynthesis, energy used in respiration, incomplete digestion, evasion and excretion.

2 a Big carnivores are at the end of food chains, so insufficient energy to support large numbers/biomass, and large territory needed to supply sufficient prey.

  b More energy lost as heat because birds and mammals maintain a constant body temperature and tend to be very active.

  c Cereal crops are young, fast-growing plants so photosynthesis produces fresh biomass quickly, which is then harvested. Mature rainforest may actually carry out more photosynthesis, but most of the carbohydrate formed is used as a substrate for respiration. Also, the productivity of crops is boosted by the application of artificial fertilizers, herbicides and insecticides.

3 a Atmospheric nitrogen (N_2) is too unreactive.

  b Lightning and nitrogen fixation by bacteria.

  c Your answer should describe these steps in eutrophication:
- increase in nitrates and phosphates
- accelerated algal growth (faster than can be eaten)
- algae die, decomposers thrive, increasing biological oxygen demand
- oxygen level falls, invertebrates and the fish that feed on them die, more decomposition, even less oxygen.

Glossary

abiotic non-living parts of an ecosystem.

absorption chemical extraction of matter into the blood.

accessory pigment molecules photosynthetic pigments that use light energy to excite electrons.

acrosome cap-like vesicle at tip of sperm, containing enzymes that penetrate egg membranes.

actin protein that, with myocin, provides contraction in muscle and other cells.

activation energy energy required to start a chemical reaction.

active transport movement of materials across a membrane against a concentration gradient.

addition mutation that leads to the addition of gene segments, or the addition of one or more amino acids in a polypeptide chain.

adenine (A) one of four bases in nucleotides.

adenoviruses viruses in which DNA is the nucleic acid; several types infect the breathing system.

adhesive forces hydrogen bonding between polar molecules.

aerobic requiring oxygen.

α-helix typical secondary protein structure held together by hydrogen bonds, forming a spiral.

allosteric activator some enzymes have a site to which the right molecules can bind, changing the shape of the active site to be compatible with its substrate, hence the reaction can proceed.

allosteric inhibitor some enzymes have an allosteric site to which the right molecules can bind. The result is to change the shape of the enzyme active site to be incompatible with its substrate molecule. Hence stopping the reaction.

alveoli air sacs in lungs where gas exchange takes place.

amino acids building blocks (monomers) of proteins. There are 20 different amino acids that only differ in the composition of their R-groups.

amnion bag-like membrane containing amniotic fluid, protects foetus from bumps.

amplify copy many times.

anabolic reaction reaction that results in building up complex substances.

anaerobic not requiring oxygen.

anaphase stage of nuclear division where chromatids move to opposite poles of cell.

antibodies molecules secreted by lymphocytes in response to specific antigens.

anticodon three bases on a tRNA molecule that join to a codon on a mRNA molecule during the translation phase of protein manufacture.

aortic bodies peripheral chemoreceptors on the aortic arch just above the heart which detect changes in blood CO_2 and pH.

appendix organ (branching off caecum) that contains cellulose-digesting micro-organisms in herbivores, such as rabbits.

apoplast pathway pathway through a plant's cell walls and intercellular spaces.

arteries blood vessels carrying blood away from the heart, towards the capillaries.

assay quantitative analysis of a substance.

assimilation build up of products of digestion into whatever substance the body requires.

arterioles blood vessels found between the arteries and capillaries.

ATP (adenosine triphosphate) energy stored in a high energy bond between adenosine diphosphate and an inorganic phosphate group.

ATP synthetase enzyme used to make ATP.

atrium chamber where blood enters the heart.

atrial systole stage in cardiac cycle when atria contract, pushing blood into ventricles.

atrioventricular node (AVN) picks up signal from SAN, relaying it through the ventricular septum and bundle of His.

atrioventricular valves one-way valves between atria and ventricles in the heart.

autotrophs organisms, such as green plants, that manufacture their necessary materials from simple components, such as water, carbon dioxide, mineral salts and light energy.

autotrophic nutrition building up organic molecules from simple inorganic molecules.

biconcave disc shape of a red blood cell.

bilayer double layer of phospholipid molecules, forming the basis of plasma membranes.

biological catalysts enzymes that that can control the rate at which chemical reactions take place.

biosphere all living organisms on Earth, together with their non-living environment.

biotic living parts of an ecosystem.

blastocyst hollow ball of unspecialised cells resulting from mitotic division of zygote.

Bohr effect when haemoglobin releases oxygen more readily in the presence of carbon dioxide

β-pleated sheet secondary protein structure caused by hydrogen bonding between adjacent polypeptide chains and forming a corrugated structure.

Brunner's glands (in duodenum) release alkaline mucus secretion to protect wall of small intestine and help neutralise stomach acid.

brush border microvilli on epithelium surface.

buffer absorbs hydrogen ions, maintaining pH.

bulk transport movement of large amounts of materials across a cell surface membrane.

bundle of His specialised muscle fibres, from which electrical stimulation spreads across ventricle walls through Purkinje fibres.

caecum first section of the large intestine.

capillaries blood vessels with walls, one cell thick.

carbonic anhydrase enzyme that converts carbon dioxide and water into carbonic acid (H_2CO_3).

carboxyl -COOH group in a molecule.

carotenoids photosynthetic pigments that absorb orange light; xanthophylls absorb yellow light.

carotid bodies chemoreceptors in the carotid artery walls that detect changes in blood CO_2 and pH.

carotid sinus swelling in carotid artery that monitors blood pressure.

carrier proteins proteins embedded in cell surface membrane that help specific polar molecules or ions to diffuse through the membrane.

carpel female parts of a flower.

Casparian strip band of suberin running round walls of endodermal cells.

catabolic reaction reaction that breaks down complex substances into simpler molecules.

cDNA 'complementary' or 'copy' DNA produced from mRNA by using reverse transcriptase.

cell division division of the cytoplasm and the nucleus into two, usually by mitosis.

cell fractionation separating the different cellular structures, ready for further analysis.

cell membrane or cell surface membrane phospholipid bilayer (7nm) that controls what enters and leaves the cell.

cell sap solution of salts, sugars and organic acids found inside of plant vacuoles.

cellulose microfibrils β-glucose units form cellulose, bonded together to form long fibres.

cell walls extra cellulose coatings surrounding plant cells and giving them extra strength.

centrioles two bundles of microfilaments that form the cell's cytoskeleton, and the spindle during division (absent from plant cells).

channel proteins transport specific molecules or ions across a membrane, by forming a water-filled tube which helps diffusion.

chemautotrophs autotrophs that use energy from chemical reactions to build organic molecules.

chemoreceptors cells sensitive to chemical change.

chemosensors cells sensitive to chemical change.

chiasmata points at which chromatids attach during crossing over.

chlorophyll a photosynthetic pigment, forming the reaction centre in a photosystem and absorbs yellow-green light.

chlorophyll b photosynthetic pigment that absorbs blue-green light.

chloroplasts organelles where the light-dependent reactions of photosynthesis take place.

cholesterol steroid with a hydrophilic head and hydrophobic tail, found in plasma membranes giving them extra strength (but less fluidity).

chorionic villi provide large placental surface in contact with capillaries of endometrium.

chromatin DNA molecules in nucleus of cells that are not dividing. It stains as a dark mass of material but without visible chromosomes.

cilia fine cytoplasmic threads projecting from the cell surface. Capable of movement and are similar in structure to flagella (only shorter).

cisternae sacs of endoplasmic reticulum.

cistron a gene.

closed blood system blood circulates round, remaining inside blood vessels.

coenzymes organic molecules that often contain vitamins, which bind loosely with the enzyme molecule while the reaction takes place.

cofactors some enzymes need extra, non-protein components to function properly.

cohesion-tension mechanism theory to explain movement of water from roots to leaves.

cohesive forces hydrogen bonding between water molecules cause them to associate and give water a high surface tension.

colloid water can keep large molecules, with charged areas on their surface, apart. This stops the particles from settling out of solution.

competitive reversible inhibitors molecules similar to an enzyme's substrate compete for enzyme active site. This slows the overall reaction down.

community all living organisms in an ecosystem.

concentration gradient the change in concentration of liquid or gas particles between regions.

condensation reaction that results in the removal of a water molecule. A hydroxyl group (-OH) on one molecule bonds to a hydrogen atom on another.

consumers heterotrophs that make up all of a food chain apart from the producers and decomposers.

corpus luteum formed from empty Graafian follicle after ovulation, secretes progesterone.

cortex layer between epidermis and vascular tissue.

Cowper's gland secretes clear fluid to clean urethra prior to ejaculation.

cristae infolded inner membranes of mitochondria.

crossing over where chromatids break and rejoin during nuclear division, exchanging genes and increasing genetic variation.

cyclic photophosphorylation cycle in photosynthesis, involving PSI, which captures enough light energy to excite electrons to a high enough level to produce molecules of ATP).

cystic fibrosis genetic disorder, which produces a faulty channel protein and thick mucus collects in the lungs and digestive system.

cystic fibrosis transmembrane regulator (CFTR) channel protein that transports chloride ions out of epithelial cells into mucus in lungs and digestive tract.

cytoplasm contents of a cell except for nucleus.

cytosine (C) one of four bases in nucleotides.

deamination removal of amine group from excess amino acids in the liver.

decarboxylated removal of carbon dioxide.

decomposers micro-organisms that break down organic compounds in dead material.

deletion mutation that leads to a loss of genes or one or more amino acids in a polypeptide chain.

denatured loss of a protein's three-dimensional structure.

deoxyribose pentose sugar found in DNA.

detritivores animals that feed on pieces of partly broken down plant or animal tissue.

digestion mechanical and chemical breakdown of large, insoluble molecules.

differential centrifugation use of a variable speed, centrifuge to separate the structures and components of cells.

differentially permeable allows the passage of some particles but not others.

differentiated cells altered to perform different functions within an organism.

diffusion movement of particles in liquids and gases from regions of high to low concentrations.

dinucleotide result of a condensation reaction between two nucleotides.

dioecious plants in which male and female flowers are on different plants, to avoid self-fertilisation.

dipeptide two amino acid molecules joined by a peptide bond.

disaccharides two sugar molecules (monomers), joined by a condensation reaction.

dissociation curve graph showing how readily a respiratory pigment releases oxygen in varying conditions.

DNA (deoxyribonucleic acid) substance that forms genes and carries the code for inheritance.

DNA ligase enzyme that joins sections of DNA. Used in genetic engineering and DNA repair.

double circulation circulatory system composed of a separate pulmonary and systemic circulation.

double fertilisation (in flowers) male gamete fuses with egg cell to form a diploid embryo and another fuses with the central cell to form triploid endosperm.

double helix (in DNA) two polynucleotides, running in opposite directions, joined at their organic bases and twisted round each other in a double spiral.

duplication a section of chromosome replicates so that a set of genes is repeated.

duodenum first section of small intestine where digestion is completed.

ecosystem ecological unit including all organisms in an area together with their abiotic environment.

electron acceptor molecules chemicals that are able to accept electrons from other substances.

electron transport chain series of compounds, which pass electrons at a high energy level, from one substance to another in a series of redox reactions, gradually releasing energy.

emulsification breaking lipids up into droplets.

endocytosis phagocytosis and pinocytosis.

endodermis ring of cells between root cortex and vascular tissue in the middle.

endopeptidases protein-digesting enzymes that act by cutting peptide chains along their length.

endoplasmic reticulum tubular channels, often expanded into cisternae, together with more or less flattened vesicles, bounded by single membranes.

endothelium epithelium lining inside of heart, blood vessels and lymph vessels.

end-product inhibition in enzyme-controlled reactions, the end products from one enzyme reaction can switch off the start of the process by entering an allosteric site as an inhibitor.

enzyme inhibitors slows down enzyme-catalysed reactions.

enzymes protein molecules that can control the rate at which chemical reactions take place.

eosinophils leucocyte involved in allergic response.

epidermis outer, protective layer of cells.

epididymis long, coiled tube at the side of the testes, where sperm are stored.

erythrocytes red blood cells.

ester bond bond formed by a condensation reaction between the hydrogen in the hydroxyl group of a glycerol molecule and the hydroxyl group in the carboxyl (-COOH) group of a fatty acid.

esterification process of bonding a fatty acid molecule to a glycerol molecule by condensation. The resulting bond is an ester bond.

eukaryotic cells cells contain a nucleus surrounded by a nuclear envelope. The genetic material is arranged on more than one chromosome.

eutrophication increased nitrate and phosphate levels in lakes or rivers, leading to reduced oxygen levels due to increased oxygen demand of decomposers.

exocytosis passage of substances from the inside of a cell to the outside, within a vacuole.

exopeptidases protein-digesting enzymes that cut amino acids off the ends of peptide chains.

expiration breathing out.

expiratory reserve volume extra air that could be exhaled in addition to tidal volume.

extracellular digestion digestion outside of an organism's cells.

eyepiece graticule scale placed in a microscope eyepiece lens and used for specimen measurement.

eyepiece lens microscope lens that you look down.

facilitated diffusion diffusion through a plasma membrane, using channel and carrier proteins.

factor VIII human blood factor needed for clotting.

fatty acids molecules with a carboxyl group (-COOH) at one end and a hydrophobic tail carbon backbone at the other end.

Glossary

fermenter vat in which conditions are optimized for microbial growth.

fibrinogen blood protein, involved in clotting.

fibrous proteins proteins form long chains, which may form hydrogen bonds with adjacent fibres.

Fick's law rate of diffusion across a membrane is directly proportional to the surface area and the difference in concentration on either side. It is inversely proportional to the membrane thickness.

filament stalk that supports the anther.

flagella fine threads, projecting from a cell and have lashing movements. They consist of 11 microtubules.

fluid mosaic model components of a cell surface membrane are constantly moving inside a phospholipid bilayer.

FSH follicle stimulating hormone, secreted by pituitary.

gamete a sex cell (egg, sperm or ovum).

gel electrophoresis technique using electrical charge to separate DNA fragments by length.

gene section of DNA containing a triplet code instruction to ribosomes and sufficient triplet codes to manufacture a specific polypeptide molecule.

gene probe section of labelled (for example, a radioactive label) nucleic acid used to locate a particular gene.

generative nucleus forms two gametes in a pollen tube.

gene therapy treating a genetic disease by inserting a working gene into the patient's cells to take over from the defective gene.

genetically identical clones; individuals with identical DNA.

genetic marker gene with an easily recognisable effect, such as resistance to an antibiotic, used in genetic engineering to label transgenic organisms.

genetic engineering artificial alteration of the genetic content of cells (not selective breeding).

globular proteins three-dimensional proteins with a highly specific shape. Hydrophilic chains project outwards and hydrophobic chains fold inwards, making them soluble.

glycerol form of alcohol with three hydroxyl (-OH) groups in the molecule.

glycolipids phospholipid with a branched carbohydrate attached to the surface.

glycolysis first series of reactions in respiration. A molecule of glucose is split into a three-carbon compound, which is then oxidised to pyruvate.

glycoproteins fibrous or globular protein with a branched carbohydrate attached to the surface.

glycosidic bond bond formed between two monosaccharide sub-units. A condensation reaction resulting in the loss of a water molecule.

Golgi apparatus in the cytoplasm of plant and animal cells. Links carbohydrates and proteins and secretes various substances from the cell.

Golgi vesicles small, temporary vacuoles 'pinched off' from the Golgi apparatus.

Graafian follicle mature follicle that protrudes from the ovary wall just prior to ovulation.

grana stack of fluid-filled sacs found inside chloroplasts.

gross primary production energy in plant biomass in a given area in a given time, harnessed by photosynthesis.

guanine (G) one of four bases in nucleotides.

guard cells pair of cells adjacent to stomata, controlling their opening.

Haber process industrial process making ammonia.

habitat where a community of organisms lives.

haem prosthetic group in haemoglobin. Haem contains four iron ions, which can each carry one molecule of oxygen.

haemoglobin red, oxygen-carrying respiratory pigment in blood.

haemoglobinic acid formed when haemoglobin combines with hydrogen ions.

haemophilia inherited disease causing an inability to make factor VIII.

heterotrophic nutrition breaking down complex organic molecules obtained from other organisms.

holozoic nutrition heterotrophic nutrition, involving ingestion, digestion, absorption and assimilation.

homogenate mixture of different sized and different density particles in solution.

homogenisation breaking up cells in a blender or using a pestle and mortar.

human chorionic gonadotrophin embryonic hormone that stimulates corpus luteum to continue producing progesterone to maintain the endometrium during pregnancy.

Human Genome Project global project aiming to decode the genes on all 46 human chromosomes.

human placental lactogen secreted by placenta and stimulates development of mammary glands.

hydrogen bonding weak bonds formed between charged parts of molecules.

hydrolysis reaction requiring addition of water.

hydrolytic digestive enzymes, inside lysosomes.

hydrophillic water-loving.

hydrophobic water-hating.

hydroxyl -OH group as part of a molecule.

hyphae threads forming mycelium of a fungus.

ileum second section of small intestine, where absorption occurs.

implantation blastocyst attaches to wall of uterus.

incisors chisel-shaped front cutting teeth.

independent assortment either chromosome from each pair passes into a gamete during meiosis, increasing genetic variation.

induced fit theory approaching substrate molecule changes the shape of the active site in the enzyme molecule so that it can fit into the active site.

ingestion taking food into the body.

insoluble non-polar molecules will not dissolve.

inspiration breathing in.

inspiratory reserve volume extra air that could be inhaled in addition to tidal volume.

internal fertilisation fusion of gametes occurs inside the female's body.

interphase stage of cell cycle when growth, specialisation and normal functions occur.

inversion mutation caused by a section of a gene or chromosome rotating by 180^0 and rejoining, reversing the amino acid sequence.

invertase enzyme converting sucrose to fructose.

isomers molecules that have the same chemical formulae but different structural formulae.

keratin protein found in hair and nails.

kinetic energy energy of movement.

Kreb's cycle cycle of decarboxylation reactions taking place in respiration.

kymograph paper mounted on revolving drum, upon which a trace can be produced with a pen.

lactation milk production by mammary glands.

lactic acid produced during anaerobic respiration when pyruvate accepts NADH rather than being oxidatively decarboxylated.

latent heat of vaporisation water requires a lot of energy to evaporate into a gas, from a liquid.

leucocytes white blood cell.

LH luteinising hormone, secreted by pituitary.

light-dependent stage of photosynthesis first stage of photosynthesis where light is used as a source of energy and water is a source of electrons and hydrogen ions.

lignin tough, woody material formed in plants.

limiting factor certain conditions that will affect the rate of an enzyme-controlled reaction.

link reaction pyruvate molecules (from glycolysis) are decarboxylated and activated with acetyl coA.

lipid-soluble non-polar solute particles can dissolve in lipids (the solvent).

liposomes vesicle containing digestive enzymes produced by Golgi body.

locating agent chemical reacts with molecule to make it visible, generally in chromatography.

lock and key theory enzyme has an active site (the lock) into which substrate molecules (the key) fit.

lumen space inside a biological structure, such as the space through which food passes down the alimentary canal.

lymphatics system of vessels which returns tissue fluid to the blood system.

lymph tissue fluid after it has drained into the lymphatic system.

lymphocytes white cells with large nuclei and little cytoplasm, which sectrete antibodies.

lysosomes membrane-bound vacuole containing hydrolytic enzymes to digest material taken in by endocytosis and assist in the removal of defective organelles and dead cells.

macromolecules giant molecules consisting of large numbers of atoms.

macrophages any large phagocytic cells.

magnification an expression of how much an image has been enlarged, or diminished.

mass flow hypothesis explanation of how mass flow can be achieved by maintaining differences in concentration and pressure rather than by a pump.

mastication chewing.

matrix the space inside the infolded inner membrane of a mitochondrion.

medulla oblongata part of hind brain, controls heart rate, breathing and blood pressure.

meiosis type of nuclear division, halving the chromosome number, producing four haploid cells.

messenger RNA (mRNA) molecule that transcribes the information carried in DNA and takes it to the ribosomes for translation into a specific protein.

metabolism total chemistry inside an organism.

metal ions some enzymes need these cofactors to change the charge inside the active site, helping the enzyme to bond temporarily with its substrate.

metaphase stage of nuclear division where chromatids attach to spindle at equator of cell.

methylene blue typical cellular stain, colouring nuclear material blue.

microfilaments fine, cytoplasmic filaments present in most cells, usually consisting of actin.

microtubules fibrous structures associated with the spindle during nuclear division.

microvilli tiny projections on the surface of a cell that increase the cell's surface area.

middle lamella intercellular material that cements adjacent plant cells together. Based on pectate.

midrib strong, lignified vein in the centre of a leaf.

mitochondria site of much respiration in eukaryotes.

mitosis cell division resulting in two identical, diploid copies of parent cell.

molars large back teeth for chewing and grinding.

molecular formula chemical formula of a substance.

monocytes large, phagocytic leucocytes made in bone marrow.

monomers small molecular units that join to form macromolecules consisting of these repeating units.

monosaccharides simplest carbohydrates (once split, they cease to have the properties of sugars).

monounsaturated one C=C in fatty acid backbone.

mucosa inner layer of alimentary canal, secretes digestive juices or absorbs food products.

multicellular organism consisting of many cells.

mutations caused by radiation and some chemicals, there are many types ranging from small changes to changes in the number of whole chromosomes per nucleus.

mycelium body mass of a fungus made of hyphae.

myogenic muscle tissue that can contract without contractions being initiated by the nervous system.

NAD (nicotinamide adenine dinucleotide) chemical involved in redox reactions in respiration.

NADP (nicotinamide adenine dinucleotide phosphate) chemical in the light-dependent and light-independent reactions of photosynthesis.

net primary production rate that plant biomass increases by photosynthesis in a given area (gross primary production minus energy lost through respiration).

neutral even distribution of electrical charge.

neutrophils white cells with lobed nucleus and granular cytoplasm.

niche how an organism fits into its environment.

nitrifying bacteria chemotrophic bacteria that convert ammonium into nitrites and nitrates.

nitrogen fixation incorporation of atmospheric nitrogen into nitrogen-containing organic compounds.

non-cyclic phosphorylation electron is excited by PSI and energy a molecule of ATP is produced by photophosphorylation. The same electron is picked up by PSII and used, along with hydrogen ions, to reduce NADP.

non-disjunction mutation caused by homologous chromosomes not separating.

non-polar even distribution of electrical charge.

nuclear envelope double membrane surrounding the nucleus in eukaryotic cells.

nuclear pores pores in the nuclear envelope.

nucleoli small, dense bodies in the cell nucleus containing RNA and protein.

nucleotides building blocks of DNA and RNA. Each nucleotide consists of a pentose sugar, a phosphate group an an organic base.

nucleus organelle containing the chromosomes.

objective lens used to focus onto a specimen on a microscope slide.

oesophagus muscular tube from throat to stomach.

oligotrophic lakes which are low in nitrates and phosphates, generally supporting more species than eutrophic lakes.

oogenesis ovum formation.

oogonia produced by mitosis in embryo.

optimum temperature enzymes have a temperature at which they work most efficiently.

organelles sub-cellular structures within cells.

osmosis special case of diffusion through a partially permeable membrane.

ovum animal egg cell or gamete.

oxidative phosphorylation series of reactions removing hydrogen ions and electrons from compounds, releasing energy and forming water.

oxygen affinity readiness of a respiratory pigment to combine with oxygen.

oxytocin hormone produced by pituitary, which stimulates uterus to contract during childbirth.

palisade cells photosynthetic cells in leaves.

paper chromatography technique used to separate out and identify different biological molecules.

partially permeable allows the passage of some particles, but not others.

partial pressure amount of a gas, such as O_2, present in a mixture of gases.

pepsin protein-digesting enzyme in stomach.

peptide bonds bond formed between amino acids, resulting from a condensation reaction.

peristalsis wave of muscle contraction, pushing something along a tube, such as the alimentary canal.

peroxisome oxidase and catalase enzymes contained within a vacule, bound by a single membrane and used to break down hydrogen peroxide to water and oxygen.

petal part of flower used to attract insects for pollination, often brightly coloured.

phagocytes cells that engulf food or bacteria.

phagocytic vesicle vacuole containing the material taken into a cell by the process of phagocytosis.

phagocytosis through an ability to stream its cytoplasm, a cell can take in large amounts of material by engulfing it.

phospholipids molecules consisting of two fatty acid molecules and a phosphoric acid molecule, joined to a glycerol molecule by condensation.

phosphoric acid H_3PO_4. Replaces one fatty acid molecule in a triglyceride to form a phospholipid.

photo-autotrophs autotrophs that use energy from light to build up organic molecules.

photolysis splitting water, using light energy.

photophosphorylation series of electron carriers involved in photosynthesis which take an excited electron and use it to produce a molecule of ATP.

photosynthesis in green plants, the process of synthesising organic compounds from water, carbon dioxide and light energy absorbed by chlorophyll.

photosystem involved in using light energy to excite electrons in the thylakoid membranes of chloroplasts. Photosystems consist of an accessory pigment, a primary pigment and a reaction centre.

phytoplankton photosynthetic plant plankton.

pinocytic channels small infoldings of a cell surface membrane, containing extracellular materials.

pinocytosis small, pinocytic channels form on the surface membrane of a cell allowing modest amounts of materials to be taken in.

pituitary gland endocrine gland that secretes several important hormones.

plasma membrane thin phospholipid bilayer (7nm) that controls what enters and leaves the cell.

plasmids ring-shaped DNA molecule found in the cytoplasm of bacteria.

plasmodesmata fine, cytoplasmic threads that pass through plant cell walls and connect adjacent cells.

pleural membranes membranes, separating lungs from inside wall of thorax.

pleural fluid lubricating fluid between pleural membranes.

polar symmetrical distribution of charge.

polar nuclei fuse to form central cell in ovule.

polymerase chain reaction (PCR) technique for rapidly producing millions of copies of a piece of DNA for analysis.

polymers complex molecule built from repeating sub units.

polynucleotide long chain of nucleotide monomers.

Glossary

polypeptide polymer consisting of many amino acids (monomers) joined by peptide bonds formed from condensation reactions.

polysaccharides polymer consisting of many monosaccharides (monomers) joined by condensation reactions.

polyunsaturated many carbon double-bonds in a fatty acid backbone.

poor conductor water is a poor conductor of heat and therefore has insulating properties.

population group of individuals of one species.

positive feedback deviatiation from a set level causes changes that exaggerate the deviation.

potassium pumps part of the mechanism used by guard cells to control the opening of stomata.

potometer instrument for measuring water uptake.

precipitation rain or snowfall.

pressure potential pressure exerted on cell contents by the cell membrane (and cell wall).

primary endosperm formed from a fertilised central cell, which nourishes the embryo.

primary follicles primary oocyte surrounded by layer of follicle cells.

primary pigment photosynthetic pigment molecule that forms the reaction centre of a photosystem.

primary spermatocytes diploid cells produced by growth of spermatogonia during spermatogenesis.

primary structure sequence and number of the amino acids in a polypeptide chain.

primers short DNA pieces that signal DNA polymerase to start copying.

producers autotrophs at the start of a food chain.

productivity amount of production per unit time.

progesterone female hormone secreted by the corpus luteum during the second half of menstrual cycle and the placenta during pregnancy.

prokaryotic cell nucleus is not surrounded by a nuclear envelope and the genetic material is arranged on one chromosome.

prolactin made by pituitary to maintain milk ducts.

prophase stage of nuclear division where chromosomes shorten and become visible, revealing chromatids.

prostate adds alkaline chemicals and clotting agent to semen.

prosthetic group group that is not made of amino acids, but associates with a particular three-dimensional protein.

prosthetic groups coenzymes that bond permanently with an enzyme molecule, helping it to function as a biological catalyst.

prostoglandins hormones that stimulate peristaltic contraction of smooth muscle.

protandry male flower parts mature first, to avoid self-fertilisation.

protocysts eukaryotic organisms, which includes single-celled protozoans that live in the digestive systems of ruminants.

protogyny female flower parts mature first, to avoid self-fertilisation.

Purkinje fibres conduct electrical impulses through the ventricles of the heart.

purines organic bases found in nucleotides and include adenine and guanine.

pyramid of biomass diagram to represent relative biomasses of populations in a food chain.

pyramid of numbers diagram to represent relative population sizes in a food chain.

pyrimidines organic bases found in nucleotides and include thymine, cytosine and uracil.

qualitative describing qualities or properties without reference to numbers.

quantitative describing qualities or properties including numerical information.

quaternary structure protein consisting of more than one polypeptide chain formed into a precise 3-dimensional shape.

reaction centre one molecule of the photosynthetic pigment chlorophyll *a*.

recognition site specific base sequence where a restriction endonuclease cuts.

recombinant DNA DNA produced as a result of genetic engineering, typically containing genes from more than one species.

reduction division another name for meiosis, resulting in haploid cells.

replica plating making several cultures from a master culture of bacteria.

residual volume volume of air remaining in lungs after breathing out.

resolution the higher the resolution, the closer two points can lie and still be seen as separate points.

respiration series of enzyme-controlled oxidation reactions transferring energy from organic compounds to ATP.

restriction endonucleases enzymes that cut DNA at specific base sequences.

retardation factor (Rf) In chromatography, every compound has its own Rf value in a particular solvent and is calculated by dividing the distance travelled by the solute, by the distance moved by the solvent front.

reverse transcriptase enzyme that synthesises a strand cDNA from an mRNA sequence.

R-group funcional group in an amino acid.

ribonucleic acid single-stranded nucleic acid involved in protein synthesis.

ribose pentose sugar found in RNA.

ribosomes site of protein manufacture.

ringing experiments determines route taken by different materials through a plant.

RNA (ribonucleic acid) single-stranded nucleic acid, involved in protein manufacture.

root hairs outgrowths of the epidermis of root cells.

root nodules swellings on roots of leguminous plants containing nitrogen-fixing bacteria.

root pressure force generated by roots, which pushes water and dissolved minerals up the stem.

RuBisco (ribulose bisphosphate carboxylase) enzyme used to join carbon dioxide to RuBP.

RuBP (ribulose bisphosphate) unstable compound formed into glycerate-3-phosphate (GP).

ruminants herbivores with rumens, such as sheep.

saprobiontic nutrition gaining nutrients from dead or decaying organic matter.

saprophytic nutrition gaining nutrients from dead or decaying organic matter.

saturated all the fatty acid bonds are satisfied.

scanning electron microscope electron beams are reflected off the surface of a specimen in a vacuum to give a high resolution image.

scolex head end of a tapeworm.

secondary oocyte produced after meiosis I of the primary oocyte.

secondary spermatocytes products of first meiotic division of primary spermatocytes during spermatogenesis.

secondary structure hydrogen and other bonds hold together sections of polypeptide chains, or adjacent chains, in an a-helix or b-pleated sheet.

sediment highest-density particles that form a layer at the bottom of a centrifuge tube after spinning.

selective barrier certain particles can move through a plasma membrane while others cannot.

semi-lunar valves one-way valves in arteries.

seminal vesicles adds fructose to semen, to provide sperm with energy for swimming.

seminiferous tubules in testes, where spermatogenesis occurs.

semi-permeable allows the passage of some particles, but not others.

sepal structure that protects the flower bud.

Sertoli cells provide nourishment for developing sperm cells during spermatogenesis.

sieve plate perforated end walls of sieve tubes.

sieve tubes part of phloem that transports sugars.

sink place in a plant or animal where substances are unloaded from a transport system.

sino-atrial node (SAN) the heart's pacemaker.

sodium-potassium pump form of active transport.

solute potential measure of the concentration of solutes in a solution on the water potential of the solution. This always has a negative value.

solvent front distance a solvent has moved through a chromatography medium.

solvent because water is polar, other polar molecules and ions will easily dissolve in it.

source place in a plant or animal where substances are loaded into a transport system.

specific heat capacity water requires a great deal of energy to raise its temperature and a lot of energy needs to be lost for its temperature to fall.

spermatids haploid cells produced by the second meiotic division of secondary spermatocytes.

spermatogenesis sperm production.

spermatogonia produced by mitotic division of germinal epithelium in seminiferous tubules.

spirometer instrument for measuring volume of breathed air and oxygen consumption.

squamous epithelium simple epithelium made of thin, flat cells that fit together like crazy paving.

stage micrometer slide microscope slide that contains a scale used for specimen measurement in association with an eyepiece graticule.

stamen male parts of a flower.

sticky ends uneven ends of DNA produced when a restriction endonuclease cuts two strands at slightly different positions.

stomata pores through which a plant loses water to the air and exchanges gases.

stop codons instruction to a ribosome telling it to stop the manufacture of a specific polypeptide.

strong triple covalent bond very stable bond, such as that between pairs of nitrogen atoms.

stroma fluid inside the double-membrane of a chloroplast.

structural formula diagram, which plots the position of each atom within a molecule.

subclavian veins veins in the neck where tissue fluid returns to the blood system.

suberin waterproof material found in cell walls.

submucosa layer under the inner mucosa of the alimentary canal, containing blood vessels and nerves.

supernatant solution that remains above the sediment in a centrifuge tube after spinning.

surface tension water tends to draw its molecules into the smallest possible volume as a result of hydrogen bonding.

surficant liquid secreted by epithelium in lungs which prevents sticking.

symbiosis nutritional relationship between two organisms of different species.

symplast pathway route through plant's cytoplasm.

tandem repeats non-coding repeating base sequences used in DNA fingerprinting.

telophase stage of nuclear division where new nuclei and new cells are formed.

tertiary structure three-dimensional shape and structure of a protein chain.

testosterone male sex hormone.

thylakoids series of flattened, fluid-filled sacs inside a chloroplast.

thymine (T) one of four bases in nucleotides.

tidal volume volume of air that moves in and out of lungs during a normal breathing cycle.

tissue fluid fluid surrounding cells in the body.

tonoplast membrane surrounding a vacuole.

total lung capacity maximum amount of air that a person's lungs can hold.

transcription process by which the genetic code in DNA is copied to produce mRNA.

transfer RNA (tRNA) carries specific amino acids to the ribosomes and is involved in the transcription process of protein synthesis.

transgenic organism contains a gene from another species.

translation joining amino acids at ribosomes to make a polypeptide chain, using the mRNA.

translocation section of a chromosome breaking off and attaching to another chromosome.

transmission electron microscope electron beams pass through a specimen in a vacuum to give a high resolution image.

transpiration loss of water from a plant.

transpiration stream movement of water through a plant, into roots, up stem and out of leaves.

triglyceride molecule resulting from the condensation of glycerol and three fatty acid molecules.

triose phosphate simple triose sugar formed in photosynthesis.

triplet code trio of bases that codes for one amino acid or an instruction code to a ribosome.

trophic level feeding level in a food chain or web.

tube nucleus directs growth of pollen tube.

tubulin hollow fibres that form a cellular skeleton to which other organelles can attach.

tunica externa external wall of arteries and veins.

tunica media middle layer of artery and vein walls.

turgor pressure pressure that the contents of a plant cell exerts on the cell wall.

ultrastructure structure at the molecular or electron microscope level.

umbilical artery carries blood from embryo/foetus to placenta.

umbilical cord connects embryo/foetus to placenta.

umbilical vein carries blood from placenta to embryo/foetus.

unicellular simple, one-celled organism.

unsaturated not all carbon atoms in a fatty acid backbone are satisfied with hydrogen. The result is a carbon double-bond, which causes a kink in the fatty acid tail.

uracil (U) base that replaces thymine in RNA.

ureters pair of tubes that carry urine from kidneys to bladder.

urethra tube which carries urine and semen out of the body.

vacuole fluid-filled space inside the cytoplasm bound by a single membrane (tonoplast).

vascular bundles groups of phloem and xylem.

vas deferens tube carrying sperm from testes to penis.

vector used to carry genes from one organism into another, for example bacterial plasmids.

vegetative propagation asexual reproduction in plants; produces clones of parent plant.

veins blood vessels carrying blood towards the heart, away from the capillaries.

ventilation rate number of complete breathing cycles per minute.

ventricle chamber from which blood is pumped from the heart.

ventricular diastole stage in cardiac cycle when atria and ventricles relax.

ventricular systole stage in cardiac cycle when ventricles contract, pushing blood into arteries.

venules blood vessels linking capillaries to veins.

vessel elements xylem tubes.

vital capacity maximum usable amount of air lungs can hold (total lung capacity minus residual volume).

water potential the ability of water molecules to move. Pure water has a water potential of zero and solutions are given a negative value.

water-soluble particles of a solute can dissolve in water (the solvent). The solute particles are generally polar in nature.

xerophytes plants adapted to dry environments.

xylem parenchyma cells packing tissue in xylem.

zooplankton animal plankton.

zygote cell formed by fusion of male and female gametes in sexual reproduction.

Index

Index